Invitation to the Natural Physics of

MATTER, SPACE, and RADIATION

Invitation to the Natural Physics of

MATTER, SPACE, and RADIATION

M. SIMHONY
(Physics Section 5, The Hebrew University, Jerusalem)

World Scientific
Singapore • New Jersey • London • Hong Kong

Published by
World Scientific Publishing Co. Pte. Ltd.
P O Box 128, Farrer Road, Singapore 9128
USA office: Suite 1B, 1060 Main Street, River Edge, NJ 07661
UK office: 73 Lynton Mead, Totteridge, London N20 8DH

INVITATION TO THE NATURAL PHYSICS OF MATTER, SPACE, AND RADIATION

Copyright © 1994 by World Scientific Publishing Co. Pte. Ltd.

All rights reserved. This book, or parts thereof, may not be reproduced in any form or by any means, electronic or mechanical, including photocopying, recording or any information storage and retrieval system now known or to be invented, without written permission from the Publisher.

For photocopying of material in this volume, please pay a copying fee through the Copyright Clearance Center, Inc., 27 Congress Street, Salem, MA 01970, USA.

ISBN 981-02-1649-1

Printed in Singapore.

To the memory of my mother, Tsirel-Braha, born Rambam, 1900,
my father Simha Szymchowicz, born 1902,
and my sister Miriam, born 1928,
murdered 1942 in Nazi-occupied Poland.

To the memory of my four uncles and two aunts,
their spouses and children,
and of the three hundred other family members and close friends,
of whom nobody remained after the Nazi occupation.
And not a gravestone.

PREFACE

What is the book about?

The book deals with the basic physical structure of atomic matter, of the physical space in and around us, and with the physical nature of radiation in space and in atomic bodies. The presentation is based on the direct results of experiments, which marked an epoch in physics. These are, primarily, the Michelson-Morley Experiments (of 1887), the Rutherford Experiments (1911), and the Anderson Experiments (1932).

The direct results of the experiments contradict our everyday perceptions and concepts of matter and space. These concepts are that matter, which we are able to perceive, i.e., atomic matter, is dense and continuous, and that space, "evacuated" of atomic matter (the "vacuum" space) is empty.

One of the basic human features is that if a finding or result does not fit our concepts, expectations or beliefs, then the result is either "silenced", or bent and twisted to make it fit.

For example, Rutherford's experiments have proven that
    all atoms and atomic matter
    consist solely of nuclei and electrons,
and that
    the distances between these particles in atomic matter
    are hundred thousand times larger
than the diameters of the particles.

These results are well known and widely used in atomic physics and in physics in general. However they also prove, that
    all bodies of atomic matter,
        including the Earth and ourselves,
            are **not** dense and continuous.

All atomic bodies represent
extremely rare networks of tiny nuclear particles,
vibrating and rotating far apart from each other.
This result is not publicized but silenced, and did not influence our way of thinking about matter.

The Michelson-Morley experiments have proven that
the motion of Earth around the Sun
does not cause any winds or currents in space.
If one knows that the Earth and all our experimental apparatus
represent rare networks of nuclear particles,
positioned far apart from one another,
then it is clear to him, that these bodies
would not be able to make winds or currents
in a rare medium of tiniest nuclear particles.
Similarly, a net made of millimeter-thick cords, which are bound a hundred meters apart from one another, cannot make winds or currents in the air or water.

Therefore, the **related result** of the Michelson-Morley experiments is, that *whatever carries the waves of light in space cannot be dense and continuous*. However, to fit our belief in the emptiness of space, the results were "generalized" to say that there is no "ether", and no physical carrier of light whatsoever. This interpretation was never corrected, to say that
the motion of Earth may not be able to cause winds or currents
in the light-carrying matter in space, if the **carrier** of light
also represents a rare network of tiny nuclear particles.

The Anderson experiment proves that
the tiny nuclear particles, existing in space,
are **electrons** and **positrons**,
bound to one another by an energy
of a million electron·volts per pair.

We could then derive (in 1973) that
the network of particles which is the physical carrier
of light and of all electromagnetic radiation,
is a rare electron-positron lattice.

This lattice, the **epola**, is also the carrier of the electric, magnetic, gravitational, and inertial forces and energies.

Unfortunately, the results of the Anderson experiments were interpreted to fit the relativistic belief that matter and energy are interchangeable and can create or annihilate each other in the empty space. This can and will be easily disproved.

How much math and physics does one need to cope with the text?

Freshmen high-school algebra is used in the text, and all the derivations are explained step by step. This is also done with the background physics. Hence, no special preparation is needed. Moreover, our math is here, merely for calculations and illustrations of the physical facts, not to prove or to explain them. Therefore, if you cannot cope with it, you will still be able to comprehend the text and enjoy the enlightenment brought by the understanding of the nature of things. An understanding, which quantum physics, relativity, and "new physics" cannot and do not provide. This was well expressed by R.P.Feynman, one of the most prominent theorists, who wrote (in 1967): "I think I can safely say that nobody understands quantum mechanics".

The main power and usefulness of mathematics is that it can be adjusted to fit everything. Therefore, it is a common mistake to consider a mathematical derivation as a proof that the physics is right. Let us not forget that for 14 centuries it was mathematically proven that all celestial bodies rotate around us; and that the velocity of light in atomic bodies is higher than in the empty space. Therefore, one should not be impressed by the popular belief that if something is mathematically complicated and incomprehensible, then it must be right.

We shall prove that the opposite is true. When the physical model is the right one, then one can achieve with simple mathematics the same results which are obtained in relativity and quantum theory by using heavy mathematical canons. And with the right physical model, one also obtains the wonderful bonus of understanding.

The aim of the book
Our goal is to reach full understanding of the physics of the phenomena with which we deal (and not only the understanding of the mathematical procedures). The book explains such unexplained and even untouched problems as
> why there are inertia and gravitation, and
>> why atoms and planetary systems are stable;
>
> what the real physical reasons are
>> for the dependence of mass on velocity,
>> for the Hubble-Humason redshift,
>> for the "3K background blackbody radiation",

what makes us think of moving with velocities close to 300,000 kilometers per second, while not a single body of the Solar System (including our "Voyager") has ever been observed to exceed 70 km/s; and many others.

With the epola model of space we are able to reach full understanding of the gravitational and inertial interaction of masses, and of the electric and magnetic interaction of electric charges and magnetic moments. However, we remain with not knowing the physical nature of **mass**, **electric charge**, and **magnetic moments**. These are intrinsic properties of the nuclear particles of which atomic bodies consist. They relate to the *"nuclear physics"* and *"particles and nuclei"* branches of physics, and are out of our scope.

Who the book is aimed at
The aim of the book is to reach all intellectually curious readers who are interested in understanding the outlined topics of physics. No mathematical or technological or physics background is needed, only lots of patience, good will, and intellectual efforts, because the book deals very seriously with very serious problems in today's physics. By no means is this a "physics made easy", or "physics for poets" book, though all derivations and discussions start from what is considered as "common knowledge".

The book aims at all people, able to understand an elementary text in biology. The motto is that
> physics deals with the simplest
>> and most basic phenomena of nature,

so that *per se*, without the mathematical and other complexities, physics must be much simpler than biology.

Why one needs this book

Really, life is very good even without this book, and without knowing and understanding the natural physics of matter, space, and radiation. But there is the surrealistic "new physics" with its mathematical inventions, of the **Big Bang** created and ever exploding universe, of **"false vacuum bubbles"**, a carry-on suitcase of which can create and destroy a universe, of **anti-matters** which threaten to annihilate us by turning our bodies into calories, of **anti-gravity** (fifth forces), which will turn the motion of our ballistic missiles and spaceships unpredictable, and many other frightening fictions.

Atop of these, there are the "star wars" books and movies, and the shrinking and blowing up of kids, as well as travel with velocities close to the speed of light, after a few years of which, thanks to the relativistic **twin-effect** and **time-reversal** fictions, one returns just in time to marry the youngest cousin of his twin-brother's grandson.

All these are false bubbles, invented and inflated to enrich their authors, publishers, and producers. So if you like to be cheated (out of your own money) by the exciting false bubbles, or if you have a financial, professional, or prestige interest in them, then avoid this book, because it will bring you disillusionment and aggravation.

But if you are a publisher, a media worker, a functionary of a scientific, political, judicial, or governmental organization, with no such interest in the bubbles, and you are faced by bubblers, who request support and funding, - don't be impressed by the mathematical derivations and codes, which you do not understand (nobody does; not even the bubbler himself).

And don't be afraid that without the support, a runaway galaxy will be caught by a competitor, while you are still in office. Thus, before allocating (wasting) the millions of public money on the bubbles, as you did on the fifth force, **read this book**. Even if you studied only law, politics, or business administration, you will reach the real freedom of mind (and freedom from being manipulated), which comes with the understanding of the physical nature of things, and grants the base to rightfully recite Gershwin's "It ain't necessarily so".

How to work with this book

Though there are only two pictures in the book, and not many a formula, it is a very serious physics book, with a very serious handling of some severe problems in physics. The reading may require the assistance of a good dictionary or desk encyclopedia. Failing to absorb a word or two can make the difference in understanding.

This book was preceded by the book *"The Electron Positron Lattice Space"* (1987, 1988, 1990), and will be followed by the text *"Introduction to Natural Physics"*. Thus, whenever it says "We have shown that..." then this showing was done in the first book, and it is not sufficiently important to be repeated here. Whenever it says "We will show that...", and it is not done here, then the showing appears in the third book.

The presentation is divided into six Circles, of about three Chapters each. The Circles, and Chapters, are self-contained. We found that it is preferable to repeat a statement made in a previous circle, than to refer the reader to it. The repetition is done in a slightly more advanced form, with a slightly different approach or goal. So one should not be too angry when he feels like shouting "I heard you!". It might be a good sign that understanding was reached. If not, please feel free to write and ask all the questions which you may have.

If the reader's attitude toward understanding is similar to ours, then lots of his patience will be needed, to read not only sentence by sentence but also word by word, while making the necessary notes, and putting the appropriate question and exclamation marks. As a reward, after the passage of all the SIX CIRCLES of this "PURGATORY", the reader will reach the enlightenment of understanding, and will find that it was worth her or his efforts. So please do not give up.

M.S., July 1993

C O N T E N T S

Preface   vii

C I R C L E   O N E

Chapter 1. MATTER, SPACE, AND NATURAL PHYSICS
1.01. The Distinct Atomic and Nuclear Kinds of Matter   1
1.02. The Emptiness and Non-Continuousness of Atomic Matter   2
1.03. The Alleged Emptiness of Space and the Imaginary Ether   3
1.04. The Takeover of Physics by Mathematics   4
1.05. The Proposed Lattice Structure of Space   6
1.06. Introducing Natural Physics   7

Chapter 2. BASIC CHARACTERISTICS OF PHYSICAL BODIES
2.01. Shapes, Measures, and Sizes of Material Bodies   9
2.02. Metric Prefixes   10
2.03. The Spherical Shape Approximation   11
2.04. Quantity of Matter and the Concept of Volume   12
2.05. Evaluation of the Volumes of Bodies   13
2.06. Metric Units of Area and Volume   14

Chapter 3. WEIGHT AND MASS OF PHYSICAL BODIES
3.01. Weight, Gravity, and Quantity of Matter   17
3.02. Weight and the Buoyant Force   18
3.03. What Is Heavier, a Pound of Lead or Feathers?   19
3.04. Mass as Quantity of Matter   19
3.05. The Law of Mass Conservation   20
3.06. Do Pan Scales Directly Measure Mass?   22
3.07. Unit of Mass   22

Chapter 4. DENSITIES OF MATTER
4.01. Density of Matter in Atomic Bodies   25
4.02. Relating to a Physics Formula and Using It   26
4.03. Densities of Substances   27
4.04. The Mass-Volume-Density Connection   28

4.05. Average Density	29
4.06. Average Densities in the Solar System and in Atoms	30
4.07. Average Densities of Bodies in the Solar System	30
4.08. Average Density in Nuclear Particles	32

Chapter 5. DENSITIES AND STRUCTURE OF MATTER

5.01. The Space-Like Emptiness of Atoms	33
5.02. The Electron Volt and Other Energy Units	34
5.03. Energies of Orbital Electrons and Atoms	35
5.04. Why Are The Empty Atoms Impenetrable To One Another?	35
5.05. Can The So Empty Atoms Be Shrunk?	37
5.06. Can Atoms of Our Bodies Be Shrunk Or Blown?	39
5.07. Aggregation States of Atomic Matter	40
5.08. Does Aggregation State Affect Chemical Composition?	41
5.09. Motion in Fluids	42
5.10. Motion of Atomic Bodies Through Collectives of Nuclear Particles	43
5.11. Are there Aggregation States of Nuclear Matter?	44
5.12. Improper Generalization of Results on Any Matter	45
5.13. Deceiving Perceptions and Concepts of Matter	46

C I R C L E T W O

Chapter 6. ABOUT PHYSICS AND OTHER NATURAL SCIENCES

6.01. What is Physics?	49
6.02. Physical Phenomena	50
6.03. Basic and Complex Phenomena	50
6.04. Penetration of Physics by Mathematics	51
6.05. Interwovenness of Natural Phenomena	52
6.06. Simplicity and Complexity of Physics	53
6.07. Effects of Extraneous Requirements on Physics	54
6.08. The Principles of Natural Physics	55
6.09. Kinds of Knowledge Raised by a Natural Science	57
6.10. Infirmity of the Explanatory Work in Physics and the Bicycle Stability Syndrome	58

Chapter 7. PHYSICAL EXPLANATIONS IN PHYSICS
7.01. Are There Physical Explanations At All? 61
7.02. Why There Are No Explanations in New Physics 62
7.03. Do We Need Physical Explanations and Understanding? 63
7.04. Educational and Social Value of Physical Explanations 64
7.05. Brief History of Understanding Physical Phenomena 65
7.06. The Inability to Understand New Physics 67

Chapter 8. MATHEMATICAL AND SCHOLASTIC MODELS IN PHYSICS
8.01. Mathematical Modeling in Physics 69
8.02. Vitality of Erroneous Scholastic Models 70
8.03. Aristotle's Physics and the "Four Horse Rule" 71
8.04. Early Geocentricity and Heliocentricity 72
8.05. Practical Needs of Knowing Planetary Motions 73
8.06. The Ptolemaic Model of Planetary Motions 74
8.07. The Copernican Heliocentric Model 75
8.08. The Keplerian Model of Planetary Motion 76
8.09. Conclusions Concerning Mathematical Models 78

CIRCLE THREE

Chapter 9. ASSESSMENTS AND INTERPRETATIONS OF EXPERIMENTS
9.01. The Direct Results of the Michelson-Morley Experiment 81
9.02. Misinterpretations of the Michelson-Morley Results 82
9.03. The Rutherford Experiment 84
9.04. Unpublicized Consequences of the Rutherford Experiment 85
9.05. Physical Interpretation of the Anderson Experiments 86
9.06. Anderson's Experiment and the $E=mc^2$ Formula 87
9.07. Analogy: Creation and Annihilation of Free Ions in Crystals 88
9.08. Inability to Make or Destroy an Electron or Positron 89

Chapter 10. THE ELECTRON POSITRON LATTICE (EPOLA) STRUCTURE OF SPACE

10.01. Space and the Fields of Forces	91
10.02. The Physical Structure of Space	92
10.03. Presentation of the Epola Model of Space	93
10.04. Our Physical Concepts of Time and Dimensions	95
10.05. The Physical Meaning of Einstein's $E=mc^2$ Formula	96
10.06. The EPOLA Structure of Space	99
10.07. Reluctance to accept the epola structure of space	100
10.08. Lattice Structure of Sodium Chloride Crystals	101
10.09. The Sodium Chloride Lattice as an Analog to the Epola	103
10.10. Calculation of the Electrostatic Attraction Energy in the Sodium Chloride and Alike fcc Lattices	104
10.11. Calculation of the Short Range Repulsion Energy in the Sodium Chloride and Alike Lattices	106
10.12. The Lattice Structure of the Epola	107
10.13. Limits for the Analogy Between the Epola and the Sodium Chloride Lattice	109

CIRCLE FOUR

Chapter 11. FUNDAMENTAL PHYSICAL INTERACTIONS AND THE INTERACTION-CARRYING SPACE

11.01. Interactions Presently Considered Fundamental	113
11.02. Ranges of Physical Interactions	114
11.03. Interactions and Their Forces	115
11.04. The Interaction-Carrying Space and Fields	116
11.05. The Inertial Interaction of Two Atomic Bodies	118
11.06. The Fundamental Character of the Inertial Interaction	119
11.07. Epola Deformation Caused By a Guest Particle	120
11.08. Epola Vibrations Caused by a Moving Guest Particle	120
11.09. Accompanying Epola Wave Around a Moving Particle	121
11.10. The Physical Nature of Inertia	123
11.11. Newton's Law of Gravitation	125
11.12. The "newton" Unit of Force	126
11.13. Calculation of Gravitational Forces	126
11.14. On the Discovery of Neptune and Pluto	127
11.15. The Physical Nature of the Gravitational Interaction	128

Chapter 12. THE ELECTROMAGNETIC INTERACTION AND THE
    ELECTRO-MAGNETO-GRAVITATIONAL (EMG) FIELD
12.01. Charge and Magnetic Moment Densities in Bodies   131
12.02. The Spread of the Electromagnetic Interaction   133
12.03. The Electrostatic (Coulombic) Interaction   134
12.04. Comparison of Electrostatic and Gravitational Forces 135
12.05. The Physical Nature of the Electrostatic Interaction 137
12.06. The Magnetic Interaction of Electric Currents   138
12.07. Spin and Intrinsic Magnetic Moments of Nuclear Particles 140
12.08. Magnetism of Currents and Intrinsic Magnetic Moments 141
12.09. Stability Condition for a System of Bodies   143
12.10. The Short Range Repulsive Interaction
       In Molecules and Lattices   144
12.11. Inertial and Magnetic Origins
       of the Short Range Repulsion   145
12.12. The "Pauli Force" Deception   146
12.13. Summary: Fundamental Interactions and the Electro-
       Magneto-Gravitational (EMG) Field in the Epola   148

CIRCLE FIVE

Chapter 13. BULK DEFORMATION WAVES IN BODIES
13.01. The Thermal Motion of Constituent Particles in Solids 151
13.02. Mean Energies of Thermal Motion   153
13.03. Conditions for the Formation of Bulk Deformation Waves 153
13.04. Half-Wave Deformation Clusters of Bulk Waves   154
13.05. Excess Particles and Energies
       in Bulk Deformation Waves   156
13.06. Energy Transfer by Wave Clusters
       and by Single Particles, or in Quanta   157
13.07. Physical Derivation of Planck's Postulated Law   158
13.08. The Phonon-Photon Analogy   159
13.09. Diverseness of Phonons   160
13.10. The Particle Features of Phonons and Photons   162
13.11. The "Particle-Wave Duality Principle" Deception   163
13.12. Nature Behind the Particle-Wave "Duality"   164

Chapter 14. VELOCITIES OF BULK DEFORMATION WAVES AND THE "MASS-ENERGY EQUIVALENCE" DECEPTION

- 14.01. Velocities of Molecules and Sound in Gases — 167
- 14.02. Propagation Mechanisms of Bulk Waves in Various Media — 168
- 14.03. Velocity of Bulk Waves and Sound in Rocksalt Crystals — 169
- 14.04. Velocities of Sound in NaCl Single Crystals — 171
- 14.05. Velocity of Bulk Deformation Waves in the Epola and the velocity of light — 172
- 14.06. Energy of Freeing and Capture of Electron Positron Pairs — 173
- 14.07. Mass-Energy "Equivalence" in Ionic fcc Crystals — 174
- 14.08. The Mass-Energy Equivalence Deception — 176

Chapter 15. SPECTRAL COMPOSITION AND NATURE OF LIGHT

- 15.01. The Spectral Composition of Light — 177
- 15.02. Continuous and Line Spectra of Light — 178
- 15.03. Interference of Light Waves and Interference Spectrometry — 179
- 15.04. Wavelengths and Frequencies of Visible Light — 181
- 15.05. Early Theories of Light — 181
- 15.06. The Electromagnetic Theory of Light — 183
- 15.07. Planck's Quantum Postulate — 184
- 15.08. Einstein's Theory of the Photoelectric Effect — 185
- 15.09. Questions on the Nature of Photons and Light — 186

Chapter 16. STABILITY AND RADIATION OF ATOMS

- 16.01. Balance of Forces and Energies in the Hydrogen Atom — 189
- 16.02. Calculation of the Orbital Velocity in the Hydrogen Atom — 191
- 16.03. Calculation of the Radius of the Hydrogen Atom — 192
- 16.04. Deficiencies in Rutherford's Planetary Model of Atoms — 192
- 16.05. Bohr's Orbits and Theory of the Hydrogen Atom — 194
- 16.06. Emission and Absorption of Light by Hydrogen Atoms — 195
- 16.07. Combined "Bohr-De Broglie" Condition for Orbit Stability — 196
- 16.08. Physical Explanation of the Stability of Atomic Orbits — 198
- 16.09. Spectra and Stability of Atoms Heavier Than Hydrogen — 199
- 16.10. Superposition of Waves — 201
- 16.11. Merging and Interference of Waves — 202

16.12. The Pauli Exclusion Principle — 204
16.13. On Achievements and Misguidings
of Quantum Mechanics — 205

CIRCLE SIX

Chapter 17. VELOCITIES OF ATOMIC BODIES AND OF NUCLEAR PARTICLES IN THE EPOLA

17.01. Velocity Limits of Atomic Bodies in the Epola Space — 207
17.02. Einstein's Formula for the Dependence
of Mass on Velocity — 208
17.03. Effects of Einstein's Dismissal
of Newton's Concept of Mass — 210
17.04. Physical Reasons for the Dependence
of Mass on Velocity — 211
17.05. Formulae for the Photon Energy, Momentum, and Mass
in Accompanying Waves — 213
17.06. The Effective Mass and Momenta in the Sublumic Motion
of a Free Electron in the Epola — 215
17.07. Calculation of the Effective Mass — 216
17.08. Lumic and Superlumic Motion of Nuclear Particles — 218

Chapter 18. FREQUENCY EFFECTS AND VELOCITIES IN WAVE PROPAGATION

18.01. Wave Intensity and Quantum Energy — 221
18.02. The Principle of Frequency Invariance — 222
18.03. The Doppler Effect — 223
18.04. Invariance of Velocities of Light and Sound
in the Doppler Effect — 225
18.05. Physics of Frequency Changes in the Doppler Effect
and Legality of their Calculation by Velocity Addition — 227
18.06. Frequency Changes in Waves due to Absorption,
Re-Emission, and Reflection — 228
18.07. Linear and Non-Linear Absorption — 230
18.08. Wave- and Propagation Velocities of Absorbable Waves — 230
18.09. Propagation Velocities of Light in Atomic Bodies — 232
18.10. Dispersion of Absorbable Waves — 233

Chapter 19. ASTROPHYSICAL ASPECTS OF THE EPOLA STRUCTURE OF SPACE

19.01. The Gravitational or Einstein Redshift — 235
19.02. Inadequateness of Relativistic Explanations of Gravitational Redshifts and Bending Light — 236
19.03. The Mirage Model for the Gravitational Bending of Light — 238
19.04. The Three Tests of Relativity — 238
19.05. The Orbit Adjustment Redshift — 239
19.06. The 3K Blackbody Radiation and the Epola Temperature — 240
19.07. The Field Temperature Redshift — 241
19.08. Temperature and Impurity Effects in the Cosmic Epola — 243

Chapter 20. COSMOLOGICAL ASPECTS OF THE EPOLA STRUCTURE OF SPACE

20.01. The Hubble-Humason Redshift and the Big Bang — 245
20.02. Non-Universality of the Vacuum Light Velocity — 247
20.03. Gravitational Presentation of the Hubble Redshift — 248
20.04. The Non-Constancy of the Hubble Constant — 249
20.05. Physical Explanation of the Non-Constancy of the Hubble "Constant" — 251
20.06. Dismissal of the Runaway Interpretation of the Hubble-Humason Redshift, and of the Big Bang — 253
20.07. Dismissal of the Creation Mechanism of Stars and Nuclear Matter by Gravitational Collapse — 255
20.08. Epola Collapse and Creation of Atomic Matter, Nuclear Matter, and Black Holes — 256
Author Index — 259
Subject Index — 261
Bibliography — 271

CIRCLE ONE

Chapter 1.

MATTER, SPACE, AND NATURAL PHYSICS

1.01. The Distinct Atomic and Nuclear Kinds of Matter

Terrestrial bodies which we perceive with our bare senses are always bodies of **atomic matter**, i.e., matter built of atoms, ions, and molecules. Consequently, what we consider as matter is usually the specific <u>atomic</u> <u>kind</u> of matter. However in this century, advancing physical apparatus enabled the discovery of material bodies which do not consist of atoms, and which we do not perceive with our bare senses.

First to be discovered were the **electrons**, found (by 1900) as parts of atoms, as the sole constituents of *cathode rays*, and of the *beta radiation* of radioactive materials. Then the *alpha radiation* of radioactive materials was shown to be a stream of much heavier particles. These **alpha particles** were later identified as nuclei of helium atoms.

Protons were found as constituents of "*canal rays*", and ascertained (by 1911) as the nuclei of hydrogen atoms. **Neutrons** were discovered in radioactive processes (in 1932) and established, together with protons, as the components of atomic nuclei. Working with *cosmic rays*, C.D.Anderson discovered the **positron** (in 1932) and the short-lived meson (in 1937). Thereafter came the discoveries of thousands of (mostly unstable) nuclei and nuclear particles.

All these particles, as well as cosmic nuclear dust and nuclear stars, are bodies of a distinct kind of matter. This **nuclear matter** differs from atomic matter not only because it does not consist of atoms, but because it is a quadrillion (a million of billions) times denser. Particles of nuclear matter have also additional quantitative characteristics named "*flavor*", "*charm*", "*strangeness*", etc. These characteristics have no parallels in atomic matter and are not applicable to it.

The recently discovered and still mysterious **cosmic matters: dark, gray,** and **black holes,** represent either distinct kinds of matter, or distinct **aggregation states** of nuclear matter. Their existence implies that there may be other, yet undetected kinds and states of matter.

1.02. The Emptiness and Non-Continuousness of Atomic Matter

With our bare senses we do not perceive single atoms, molecules, and ions, of which atomic bodies consist. This is why we perceive bodies of atomic matter as being DENSE and smoothly CONTINUOUS. These *natural perceptions* resulted in our *scientific concepts* of the denseness of solids and liquids, and of the continuousness of all matter, also of the air and other gases. The calculative treatment of matter is therefore conducted by the best developed "**mathematics (or physics) of continuous media**".

However **Rutherford's Experiments** (1911) showed that alpha-particles (nuclei of helium atoms) can move almost undisturbed through metal foils. This can be possible only if the atoms of these foils, and of bodies in general, are almost empty. Rutherford found that almost all the mass of an atom is concentrated in its NUCLEUS, the diameter of which is a hundred thousand times smaller than the diameter of the atom. The volume of the nucleus is therefore a quadrillion (a million billions) times smaller than the volume of the atom.

The masses and volumes of the electrons are thousands of times smaller than the masses and volumes of the atomic nuclei. Hence, the atomic electrons, which are circling or *orbiting* around the nucleus in the atom (ORBITAL ELECTRONS) have insignificant masses and volumes compared with the mass and volume of the nucleus. Therefore, only a quadrillionth of the volume of any atom is "filled" by the volume of the nucleus and the electrons of which the atom consists. The rest of the volume, thus *almost all the volume of any atom is just space*.

For comparison, the volumes of the Sun and of all planets and other bodies of our Solar System "fill" less than a trillionth (a millionth of a millionth) of its volume. Hence, the volumes of atoms are still a hundred times "emptier" than the volume of the Solar System.

Because of the "space-like" emptiness of atoms, thus also of molecules and ions, *all atomic bodies, including ourselves, are space-like empty,* too, and not dense, as perceived. Only a quadrillionth of the volumes of atomic bodies is filled by nuclear particles, which are the nuclei and electrons of their atoms. The rest of the volume, i.e., almost all the volume of atomic bodies is just space.

Atomic bodies are also not continuous but discrete; firstly, because of the graininess of their molecular, atomic, and ionic structure. Then, atomic bodies are not continuous but discrete, because *the distances between the particles of which the atoms consist are a hundred thousand times larger than the sizes of these particles.*

Dealing with phenomena, which involve the submicroscopic structure of atomic matter, one should remember that atomic bodies are not how we see or feel them. Each atomic body, including our own, is actually a network of nuclei and electrons, located far apart from one another.

1.03. The Alleged Emptiness of Space and the Imaginary Ether

The measured velocities of planets did not decrease during centuries of observation, and remain as high as they were 300 years ago. It is therefore clear that all motions of atomic bodies in space occur without resistance. For example, the velocity of Mother Earth in Her motion around the Sun was and is around 30 kilometers (almost 19 miles) per second (over 100,000 kilometers per hour).

Because we perceive and consider atomic bodies as being dense and continuous, we cannot imagine that those dense and continuous planets of enormous sizes can move with such tremendous velocities through any continuous material ambient without resistance. The obvious conclusion is therefore that space must be empty. We think, that if space were to contain any material, then this material would necessarily be continuous, and it would inevitably have to resist the motion and reduce the velocities of the moving bodies.

The absolute emptiness of space was the basic assumption in all theories derived by Sir Isaac Newton, concerning motion, inertia, gravitation, and light. He created the *corpuscular* theory of light, in which light was considered a stream of corpuscles (particles), emitted by the source, and propagating through the empty space. The corpuscular theory of light was reigning in science for 140 years, till 1825.

Newton understood that the empty space cannot carry waves, therefore he opposed the *wave* theory of light, created by Christian Huygens. However by 1825, it was finally and irreversibly proven that light represents wave processes. One would then immediately ask "WAVES OF WHAT?", or IN WHAT, because waves need an ambient, a CARRIER in which to form and propagate.

This ambient or carrier of light waves was introduced under the name of ETHER, supposed to fill up all the empty space. To carry the waves of light, the ether had to be highly elastic, and as stiff as steel. On the other hand, in order not to resist the motion of bodies in space, the ether had to be massless. Hence, the ether was physically unfeasible, it could exist only in our imagination, not in nature.

Nevertheless, the imaginary ether concept was quite fruitful in physics, especially in James Clerk Maxwell's theory of electromagnetism (1865). In Maxwell's theory, the ether was assumed as the carrier of electric and magnetic forces and FIELDS, and light was presented as a kind of **electromagnetic radiation**, as electromagnetic waves of certain frequencies. It is important to note that Maxwell's theory of the electromagnetic field, and the MAXWELL EQUATIONS, which still constitute the calculative basis of electromagnetism, were derived in a mathematical treatment of the ether as a continuous medium.

However the **Michelson-Morley Experiments** have proven (in 1887) that the motion of Earth around the Sun does not cause any "**winds**" or "**currents**" in the ether, or in whatever it is that carries the light waves, nor does this motion push or sweep the carrier. This actually means that *whatever it is that carries light waves* **cannot be continuous!** Unfortunately, being human, thus unable to *imagine* a non-continuous (discrete) ambient, let alone to accept one, physicists decided that there is no ether, and no physical carrier of light waves whatsoever. Henceforth, Einstein revived in 1905 Newton's postulation that space is empty.

1.04. The Takeover of Physics by Mathematics

Though the ether was physically unfeasible, it still was a kind of an "as if" quasi-physical model of the radiation-carrying space. When the existence of the ether, actually, of a continuous ether, was excluded by the Michelson-Morley experiments, physics remained without even this etherous shred of a physical model, and became a science in distress.

During the hundred year long frustration of *"life without ether"*, i.e., without any kind of a physical model of space, but with an absolute belief in *dense and continuous* matter and *absolutely empty* space, messianic mathematical inventions were introduced to enable calculations, which would fit the observables. Among those were, first, the assumptions that the length of a moving body shortens (*length dilation*), that in a moving body time runs slower (*time dilation*), that time can run backwards (*time reversal*), etc. These mathematical inventions contradict the physical concepts of real time and of the three real space dimensions of bodies. They do not occur in reality and do not reflect it.

Later came the "add a dimension" game, increasing the number of dimensions, needed to describe the motion of a body. It started in 1919, with Kaluza's fifth dimension, and led, in 1985, to M.G.Duff's 506 dimensions. In parallel, the competing "add a particle" game was and is still going on. In this game, one "creates" and "annihilates" non-existing "particles", just by writing or "applying" an appropriate mathematical function, an operator, or fractal. Hundreds of such imaginary particles have been "created". On paper, and in papers, of course, not in reality.

With all these inventions, it turned out that space, which was postulated to be absolutely empty (1905), must also have some properties of physical bodies. One after another, the empty space was granted the property of being deformable, then of having energy, then of being electrically charged, then magnetized, etc. If you asked, in this **"new physics"**, how can an emptiness have these properties, you would be shown mathematical derivations, which "prove" it, unequivocally, of course. If not convinced, you will be proclaimed an ignoramus, or just a fool, as in Hans Christian Andersen's story about the new clothes of the king.

The ways of apprehension and cognition, thanks to which physics was established as a science, were replaced in "new physics" by the introduction of unexplained postulates and principles, in the same way by which axioms are introduced in mathematics. All this, through disrespect, disregard, and denial of thoughtful physical definitions and laboriously derived and proven physical laws.

The previous tool of physics, named "the mathematics of physics and chemistry", or "mathematical methods of physics", became *mathematical physics*, and even *theoretical physics*. Hence, there is no more *thinking* physics, no more physical models, no more real physical theories, no more physical explanations of the nature of physical phenomena. All this was and is replaced by inventing quantum and relativistic rules of how to calculate.

1.05. The Proposed Lattice Structure of Space

The almost undisturbed motion of alpha particles through atomic bodies, disclosed by Rutherford, and the even lesser disturbance to the motion of neutrons in atomic bodies, mean that atomic bodies can move just as undisturbed through collectives of such nuclear particles. This is so because the distances between the nuclear particles, constituting an atomic body, are up to a hundred thousand times larger than the sizes of nuclear particles. Similarly, a fisherman's net, woven of millimeter thin cords, a hundred meters apart from one another, could move undisturbed through schools of sardines and herring.

Consider an *electrically neutral* collective of electrons and positrons, bound to one another at interparticle distances 50 times larger than the radii of the particles. This distance is comparable to the "medium to large" sizes of atomic nuclei. Now suppose that space around us represents such a collective, or network, or **lattice** of electrons and positrons. An atomic body, moving in this lattice, would "sweep" the nuclei and electrons of its atoms through the interparticle distances in the lattice, causing no winds or currents, only vibrations and waves.

These waves correspond to the "waves of matter" or "de Broglie waves", introduced in quantum theory by L.V.de Broglie in 1924. Similarly, a fisherman's net, of thin cord and large eyelets, does not cause currents when moving in water, but only wavelets around the nodes of the net.

Hence, the Rutherford experiment leads also to a physical explanation of the *"no ether winds"* results of the Michelson-Morley experiments. The motion of an atomic body in space makes no winds in it, because this is not the motion of a dense continuous wall in a continuous ambient, as we are used to think, but the motion of a highly "diluted" network of nuclear particles, through a lattice of the tiniest nuclear particles, positioned far apart from one another.

Neither of the atomic bodies, not even the Earth or Sun, is dense and continuous. All atomic bodies, and the space, carrying the *electromagnetic and gravitational field*, consist of nuclear particles, located far apart from one another. Atomic bodies can therefore move undisturbed through the electron positron lattice space. We shall name this space lattice **EPOLA**, for short.

The nuclear particles, constituting the atoms of the body, constantly interact with the electrons and positrons of the epola. When the atomic body rests, the interaction deforms the body and the lattice, resulting in gravitational, electric, and magnetic forces. When the body moves, the interactions result, additionally, in the observed de Broglie waves in the field, and cause the *inertia* of the body. The interactions also raise the calculated *effective* mass of nuclear particles when their velocity increases toward the velocity of light.

It is important to note, that collectives of electrically neutral nuclear particles, or particles of opposite electric charges, like electrons and positrons, would not be detected by our "regular" means, nor could they be pumped out by our best vacuum pumps.

1.06. Introducing Natural Physics

The electron positron lattice (epola) model of the structure of space and of the electro-magneto-gravitational field makes it possible to turn all working postulates and principles of quantum theory and relativity into physically derivable and explainable laws, and to provide full physical explanations of all physical phenomena occurring outside the nuclear particles. Based on this model, physics can be restored as a natural science, as natural as biology still is.

To distinguish the restored physics, or just physics, we name it "natural physics". This may read as "buttery butter", because the word *physics* itself means nature (*physis* in Greek). Nevertheless, we shall use the name of *natural physics* for as long as there is the non-natural physics, and the unfriendly mathematical takeover continues.

Fighting the mathematical takeover of physics should by no means be understood as fighting mathematics. No serious science can do without math, and physics without math would not exist. Actually, in our times, nothing in life can do without math, not even a successful household. Math is necessary also in love, if we want it to last for life. However, not as a replacement of the natural, as it became in the "new" physics.

It should be clear that mathematics is not a natural science. Its objects do not exist in nature, and the axioms, rules and laws of math are "man made", to fit the particular ideas and needs of their creators. Thus, e.g., if one does not like the geometry in which two parallel lines cannot cross one another, he is free and welcome to create a geometry in which they can. This does not make any difference to nature, because no geometrical line can anyhow exist in reality, and no real lines can be parallel, neither in the first sense nor in the second.

To mathematics, physics (including mechanics) is a tool, a source of inspiration, and a most important training ground. To natural physics, as to any other natural science, mathematics is a tool for calculations. Not a tool to prove or verify physical models, and certainly not a tool to create and annihilate particles or anything real.

To conform with reality, natural physics must make proper use also of its other tools. These are technology, which provides experimental apparatus and methods, philosophy, which provides methods of thinking and verifying observables, and the language of physics, which provides ways of communication and order. The most important tool is always the one which we "forget" to use. All developed and advancing mathematical approaches should be used in natural physics, but as calculative tools, not dictates. They should not be allowed to diminish the role and use of the other tools of physics, nor to replace them.

Natural physics is actually what one would expect of physics to be - the natural science of physics. We shall therefore write just "physics" whenever it is clear that the natural physics is meant. The non-natural, unexplainable non-physical physics, in which particles and universes are created and destroyed by writing or "applying" a "proper" mathematical function, operator, or fractal, shall be marked by its branch name, or just as "new" physics.

Chapter 2.
BASIC CHARACTERISTICS OF PHYSICAL BODIES

2.01. Shapes, Measures, and Sizes of Material Bodies

The shapes of material bodies are usually very complicated. One becomes aware of it when trying to reproduce those shapes, depict them, or just design an enclosure, a dress to fit. For these purposes, one may adequately describe the shapes by some **linear measures** of the body. These are usually characteristic lengths between specific points of the body, *cross-section diameter*s, *circumference*s, etc.

Bodies, the shapes of which are close to imaginative **ideal geometric shapes**, can be quite exactly described by very few measures. For example, a precisely machined sphere can be described by one measure: the diameter of the sphere, or the radius, which is half the diameter. Also a precisely machined cube can be described by one measure, the edge l of the cube.

The number of measures needed to describe other shapes grows with their complexity, and with the degree of accuracy requested in the description. The number is especially large for the shapes of natural bodies. Imagine how many measures have to be taken to make a sculpture of a human body, like Michelangelo's David. On the other hand, one may obtain a general (and quite exciting, or irritating) impression of a person just from her three measures, like "38, 24, 36". One may also adequately fit a dress for a person, knowing a single generalized measure, the **size**, like "size 12". The word **"size"** was adapted to denote a measure of a body, when the other measures are not significantly different, or not important for our particular purpose, or just unavailable.

Assume, e.g., that we succeeded to line-up some 50 small but still visible dust particles, and that the length of this line-up is one millimeter or 1000 microns (or micrometers, or millionths of a meter). Dividing 1000 microns by 50, we find 20 microns as the single obtainable measure of such a dust particle. We may then state that the size of the smallest bodies, still visible to our bare eyes, is approximately (or about, "~") 20 microns (~20 µm).

Very distant and very small bodies are seen blurred, with undefinable shapes. Their measures are not clear and do not decisively vary with direction. Thus, the available or relevant measures reduce to a single approximate measure, the size. Such are the quoted sizes of galaxies and stars, nebulae and clouds, comets and meteors. Such are also the quoted sizes of the smallest visible and invisible bodies, of molecules, atoms, and of nuclear particles. All these sizes were deduced without knowing the actual shapes of the bodies.

2.02. Metric Prefixes

One of the greatest advantages of the metric system is the ease and simplicity of creating and using larger and smaller units by attaching standard metric prefixes to the names of the **basic** units, like meter, gram, second, and liter. For example, 1 kilometer is thousand meters, and in symbols,

$$1 \text{ km} = 1000 \text{ m, or } 1 \text{ km} = 10^3 \text{ m.}$$

A millimeter is an example of a fractional unit; 1 millimeter is a thousandth of a meter, and in symbols,

$$1 \text{ mm} = 0.001 \text{ m, or } 1 \text{ mm} = 10^{-3} \text{ m.}$$

In the *"powers of ten"* notation of numbers, thousand is *ten to the third power*, or $10 \cdot 10 \cdot 10 = 10^3$. By the mathematical rule for this short notation, the superscript "power index" shows the amount of digits in front of the decimal point in the number. In powers of ten, a thousandth, or $1/10 \cdot 10 \cdot 10 = 0.001$, is 10^{-3}, *ten to the minus third power*. In such a short notation of decimal fractions in powers of ten, the superscript <u>negative</u> power index shows the amount of zeros behind the decimal point, increased by one. For example, a microsecond (µs) is a millionth of a second (s), thus

$$1 \text{ µs} = 0.000001 \text{ s, or } 1 \text{ µs} = 10^{-6} \text{ s.}$$

Also, 1.02 MeV (one point zero two mega electron·volt) is a million and 20 thousand electron·volt,

$$1.02 \text{ MeV} = 1{,}020{,}000 \text{ eV} = 1.02 \cdot 10^6 \text{ eV}$$

(one point zero two, times ten to the sixth power electron·volt). The prefixes and their values are listed in Table 1.

Table 1. Metric Prefixes

Prefixes Creating Smaller Units

Prefix	Symbol	Fraction of basic unit	Math Value
deci	d	a tenth	$0.1 = 10^{-1}$
centi	c	a hundredth	$0.01 = 10^{-2}$
milli	m	a thousandth	$0.001 = 10^{-3}$
micro	μ	a millionth	$0.000001 = 10^{-6}$
nano	n	a billionth	10^{-9}
pico	p	a trillionth	10^{-12}
femto	f	a quadrillionth	10^{-15}
atto	a	a quintillionth	10^{-18}

Prefixes Creating Larger Units

Prefix	Symbol	Rate of unit increase	Math Value
deka	D	ten times	10
hecto	h	hundred times	100
kilo	k	thousand -"-	$1000 = 10^3$
mega	M	million -"-	10^6
giga	G	billion -"-	10^9
tera	T	trillion -"-	10^{12}
peta	P	quadrillion -"-	10^{15}
exa	E	quintillion -"-	10^{18}

2.03. The Spherical Shape Approximation

Quite often the single approximate measure or size of a body is quoted as its diameter. However, the diameter is, by definition, the appropriate single measure for ideal imaginary **geometric** circles and spheres. Hence, listing the diameter or radius instead of the size implies that *the shape of the body is approximated as spherical.*

When we quote the diameters or radii of celestial bodies, or of atoms and nuclear particles, we actually consider their unknown shapes spherical. This *spherical shape approximation* is our best choice for all unidentified shapes, as long as it is not contradicted by more precise measurements or more specific requirements.

Our planet Earth may well be approximated as a sphere. Certainly, the particular radius, emerging from Earth's center and pointing towards the highest peak of Mt. Everest, is about 9.3 kilometers longer than the radius reaching at the Dead Sea the point of Earth's lowest depression. However this difference is only ~0.15% of Earth's equatorial radius, which is 6378 km. The polar radius is 6357 km, by 21 km, or 0.33% shorter. Earth's mean radius is 6370 km.

To depict these differences on a globe, 30 cm in equatorial diameter, the polar axis of it should be 0.6 mm shorter, and Mt. Everest should stick out 0.25 mm altogether. Hence, as rough as the Earth appears to travelers, explorers, and road builders, it is proportionally smoother and closer to an ideal sphere than, e.g., natural or artificially grown pearls. However in a most precise ball-bearing, the diameters of a ball (e.g., 5 mm) may differ from one another by about 1 micrometer or less, i.e., less than by 0.02%. Such balls may well represent our practically best achievable approach to ideal geometric spheres.

To conclude, not a single material body has the shape of a geometric sphere, but almost any material body can be considered spherical (or cubic, or ellipsoidal, etc.) **within the limits of the approximation**. However, serving well any particular purpose(s), the approximation may not be good for another purpose. For example, as much as one may be impressed by his knowledge that the Earth's surface is spherical, he must not insist on it when outlining a railroad. All our approximations and approaches have limits of applicability. We have to be aware of these limits when formulating our conclusions and laws, and before applying them, but especially - before **generalizing** them on most everything, most everywhere, as we so often like to do.

2.04. Quantity of Matter and the Concept of Volume

Our first judgement on the quantity of matter in a body is based on its <u>linear</u> measures. It is clear to us that among bodies composed of the same substance(s), the one with larger measures contains more matter. The comparison is easy when the bodies differ by one measure only, e.g., by length, while the other measures (thicknesses, widths) of the bodies are identical. Then we not only accept that the body with the larger particular measure (e.g., the longer spaghetti) contains more matter, but also that the quantity of matter in it is directly proportional to this measure; e. g., a body three times longer contains three times more matter.

The judgement becomes difficult when bodies differ by more than one measure, or by a measure which accounts for the cross-section **area** or for the **volume** of the body. For example, looking at an egg, which is shorter but thicker than its contestant (has thus a larger cross-section area), can you decide, if it contains more matter? And if you do, can you approximate, by how much? Looking at a solid spherical body, the diameter of which is thrice that of a second one, of same substance, do you *feel*, or do your "natural concepts" tell you that it contains about thirty times more matter?

All these cases lead to the connection between the physical concept of the quantity of matter in a body and the geometric concept of the volume of the body. The volume of a physical body can be defined as *the amount of space, occupied by the body*. Hence, the volume is a **spatial measure**, distinct from linear and surface measures. It is a combination of these measures, the more complicated, the more such measures are needed to define the shape of the body. While a sculptor may be well trained to estimate linear measures by eye, the comparison of volumes, let alone their valuation, would be difficult even to him.

It is clear to us that the quantity of matter in a body should be proportional to the space, occupied by the body, thus to its volume. The difficulty in evaluating the quantity of matter in a body is then in the evaluation of the volume of the body.

2.05. Evaluation of the Volumes of Bodies

The easiest and most natural way to evaluate the volume of a body is by immersing the body in a liquid which does not dissolve the substance of the body, and measuring the amount of liquid which was ousted by the body. The volume of the ousted liquid is necessarily equal to the volume of the body.

A more accurate method is to weigh the body while it is hanging down from the balance, then to immerse the hanging body in the liquid (without wetting the balance), and to weigh it again. The measured reduction in the weight is the **buoyant force** acting on the immersed body, and is equal to the weight of the liquid, ousted by the body (Archimedes' Law). Knowing the weight of the ousted liquid, one can find its volume, which yields the volume of the immersed body.

If the shape of the body (or of its parts) can be approximated by an ideal geometric shape, then the volume of the body can be evaluated using the mathematical rules and formulae derived for these shapes. For example, to calculate the volume of a parallelepiped (brick, box, room), one has to multiply its three dimensions by one another: its length l by its height h, and by the width w. To calculate the volume of a cube, one has to raise the length l of the edge of the cube to the third power, because

$$l \cdot l \cdot l = l^3 .$$

To calculate the volume of a sphere, one has to multiply the area A of the large circle of the sphere (largest cross-section area of the sphere), which is

$$A = \sim 3.14 \; R^2,$$

by four thirds (4/3) of the radius R of the sphere. Hence, the volume V of the sphere is

$$V = \sim 4 \cdot 3.14 \; R^3/3 = \sim 4.2 \; R^3 .$$

We see that in all expressions for the volume, <u>three</u> linear measures are multiplied by one another, and the units thus appear raised to the third power, or **cubed**. In formulae for the area, <u>two</u> linear measures are multiplied, and the units are raised to the second power, or **squared**.

2.06. Metric Units of Area and Volume

The metric units of area and volume are *composite units*, derived from the basic unit of length, the meter. Hence, the metric unit of area is the square meter; the metric unit of volume is the cubic meter. The units, and some of their subunits, are listed in Table 2.

Table 2. Some Metric Units

of Length	Area	Volume
meter	square meter	cubic meter
1 m	$1\ m^2 = 1m \cdot 1m$	$1\ m^3 = 1m \cdot 1m \cdot 1m$
kilometer	$1 km^2 = 1km \cdot 1km =$	$1km^3 = 1km \cdot 1km \cdot 1km =$
1km=1000m	$= 10^6\ m^2$	$= 10^9 m^3$ (a billion m^3)
decimeter	$1dm^2 = (0.1m)^2 =$	$1\ dm^3 = (0.1m)^3 =$
1dm=0.1 m	$= 0.01\ m^2$	$= 10^{-3}\ m^3 = 1$ **liter**
centimeter	$1cm^2 = (0.01m)^2 =$	$1cm^3 = (0.01m)^3$
1 cm=0.01 m	$= 10^{-4} m^2$	$= 10^{-6} m^3$ (a millionth)
millimeter	$1mm^2 = (10^{-3}m)^2 =$	$1mm^3 = (10^{-3}m)^3 =$
1 mm=0.001m	$= 10^{-6} m^2$	$= 10^{-9} m^3$ (a billionth)
micrometer	$1\mu m^2 = (10^{-6}m)^2 =$	$1\mu m^3 = (10^{-6}m)^3 =$
1 μm=10^{-6}m	$= 10^{-12}\ m^2$	$= 10^{-18}\ m^3$

Note. In squared and cubed units with a prefix, the power index (2, 3), is over the meter, but relates to the prefix as well; thus,

1 km^2 is actually 1 $(km)^2$; 1 dm^3 is 1 $(dm)^3$, etc.

Chapter 3.
WEIGHT AND MASS OF PHYSICAL BODIES

3.01. Weight, Gravity, and Quantity of Matter

A more direct judgement on the quantity of matter in a body we obtain by weighing. The weight of a body is defined as a measure of the gravitational force, with which the body is attracted to Earth. This force stretches the strings or springs on which the body hangs, compresses and bends the supports or springs beneath the body, or accelerates the fall of the body when it is not supported.

It is then clear to us that the gravitational force, acting on a body, must be the stronger the larger the quantity of matter in the body. Hence, the weight of a body should be proportional to the quantity of matter in it.

The SI **unit of force** and weight is the **newton**, symbol N. A newton is equal to the weight of a body of mass ~0.102 kg. The popular non-SI unit of force is the **kilogram-force**, or **kgf**, for short, which is the weight of a body having a mass of 1 kg. The kilogram-force is thus equal to 9.8 newtons.

The gravitational force and the weight on or above Earth's surface decrease with the remoteness of the location from the center of Earth. The force is largest at sea-level on the poles. At sea-level on the equator it is smaller by ~0.5% . This means, e.g., that a body weighing a kilogram-force on the North Pole, weighs 0.995 kgf at sea level on the equator. The gravitational force decreases with increasing elevation, and, e.g., at a height of ~6400 km, where the distance to Earth's center is *twice* the radius of Earth, the gravitational force is *four times smaller than on Earth's surface*, and so is the weight.

When the body rotates in orbit around Earth, whether in or outside a satellite, the gravitational force provides the centripetal acceleration, necessary to keep the body on that orbit. As long as the orbit is stable, the gravitational force is completely "used up" in providing the centripetal acceleration. The force cannot stretch or compress the supports of the body or of its parts, nor can it accelerate the fall of the body when it is not supported; the body is **weightless**.

We solemnly believe that the quantity of matter in a body remains unchanged when it is moved to all those places on Earth and in space. Therefore, the changing and even disappearing weight is not a direct measure of the quantity of matter. Nevertheless, by accounting for the described factors, which affect the weight, and by being alert to changing conditions and other factors, known and yet unknown (the world is full of surprizes!), one may use weighing as an appropriate way to determine the quantity of matter.

3.02. Weight and the Buoyant Force

Another problem encountered in weighing is the buoyant force in the air (and much more so in water). The buoyant force acts upwards, opposite to the gravitational force. It thus reduces the measured, or **effective** weight of bodies the more, the larger their volumes (Archimedes' Principle).

For example, when a body of volume 10 liters (or 10 cubic decimeters) is immersed in water, it pushes up or ousts a volume of water, equal to its volume. The gravitational force, acting on the ousted 10 liters of water, is 10 kgf, or 98 newtons. This force "wants" the ousted water back, down as low as possible. Hence, the gravitating water is acting on the immersed body with an upward buoyant force of 10 kgf. Thus, if the body weighs 40 kgf in air (or better yet, in a vacuum chamber) then, immersed in water, it weighs only 30 kgf.

Under *normal conditions*, a liter of air weighs ~1.3 gram-force (1.3 gf). In the air, a block of polystyrene foam of volume 30 liters, ousts 30 liters of air, which weigh 30 1.3 gf, or 40 gf. The air, ousted by the block, acts on it with a buoyant force, equal to 40 gram-force.

If the block weighs ~1 kgf in air, then this **effective weight** is reduced from the actual weight by ~40 gf. The weight of the block in a chamber where air was pumped out (in a **vacuum** chamber) would thus be ~1040 gram-force; this is the actual weight.

The buoyant force acting on a balloon, filled with a light gas, can become larger than the weight of the balloon. Then the **effective weight is negative**, and the balloon flies up. However we solemnly believe that the quantity of matter in a balloon is not reduced after filling with gas; to the contrary, it must be increased by the quantity of matter of the gas itself.

3.03. What Is Heavier, a Pound of Lead or Feathers?

This question is provocative; it hints at a common feeling of the heaviness of the piece of lead as compared with the lightness of feathers. Therefore, unsophisticated people would usually say that a pound of lead is heavier than a pound of feathers. Sophisticated people overcome the common feeling and use common sense, leading them to the conclusion that none is heavier; if both weigh the same, they must be equally heavy.

This answer seems so clear, simple and true, that to some psychologists it serves as a sign of the fine mental abilities and sanity of their patients. The question is therefore considered a useful tool in evaluating those abilities.

However, the buoyant force acting in air on the feathers is significantly larger than the one acting on the lead. When the feathers and the lead balance in the air, their _effective_ weights are equal. However the actual weight of the feathers, i.e., the gravitational attractive force, exerted by Earth on the feathers, is larger.

This can be demonstrated by reducing the air pressure around the balance with the two bodies on their pans, e.g., by pumping out some air. At reduced air pressure, the pan with the feathers goes down, showing that the "pound of feathers" is heavier.

3.04. Mass as Quantity of Matter

The use of the term "mass" as an equivalent for the quantity of matter was introduced by Sir Isaac Newton. In his book "Mathematical Principles of Natural Philosophy" (1687) he stated that "Mass is quantity of matter" ("*Massa est quantitum materiae*").

To the masses of bodies as quantities of matter in them, Newton attributed the phenomena of inertia and gravitation. He defined these phenomena and the forces, acting on the participating bodies, and he established that the gravitational and inertial forces

acting on bodies, or exerted by them,
are proportional to the masses of the bodies.

Newton also established that the accelerations acquired by bodies under the action of forces, are proportional to these forces, and
inversely proportional to the masses of the bodies.
Hence, one can determine the mass of a free body by measuring the force and the acceleration which the body acquired due to this force.

The mass of a body, determined from the force and acceleration, is the "inertial mass", while the mass determined by weighing is the "gravitational mass". Newton stated that these "two masses" are identical. Actually, a body has only one mass, so that the inertial and gravitational **methods of finding** the mass, as well as any other method, should yield **one** value, within the accuracy limits of the experiments. If not, then one of the methods (or some, or all) cannot be applied!

Such an inapplicable method would be, e.g., the weighing of a body in an orbiting satellite. There the body is weightless, because the action of the gravitational force is "used up" on providing the centripetal acceleration, necessary to keep the body in orbit. Except for this, all gravitational and inertial measurements, performed until now, do yield identical values of the "two masses" for bodies of atomic matter. However, we cannot be sure that it will be so also with masses of nuclear particles, moving with velocities close to the speed of light, should we succeed in performing the experiments.

3.05 The Law of Mass Conservation

The basic feature of the mass as quantity of matter in a body is its constancy. To change the mass, one would have to cut, melt or burn away a part of the body, or adhere other bodies to it, i.e., to change the body itself. But even then *the mass is conserved*, meaning that the sum of the masses of all bodies before the changes is equal to the sum of all masses of all bodies after the changes, including the masses of all products of burning, or of other chemical reactions involved.

This is the content of the Law of Mass Conservation, established by Antoine Laurent Lavoisier (1743-1794). The Law is one of the most important laws of physics and chemistry. It is also the most laboriously and scrupulously established and proven law of the natural sciences.

With great sorrow we mention the ease with which a French revolutionary court disposed of Lavoisier, sending him to the guillotine for just a misdeed. Against the voices of defence, the verdict stated that "the revolution does not need scientists". With similar ease, though bloodlessly, the rebellious "new physics" disposed of his Law, just because mathematical approaches do not need stable experimentally established pillars on their way.

Mass Conservation and Indestructibility was the most stable and reliable pillar of physics and chemistry until Einstein interpreted his $E=mc^2$ equation, as a proof that mass and energy are *equivalent* to one another and do thus turn into one another. We will show later that this equivalence is purely mathematical, because no transfer of mass into energy, or energy into mass can occur in reality. Any apparent turning of mass into energy and of energy into mass is a result of not taking into account the masses of **all** participants of the reaction, also the masses of the epola particles, freed of their bonds, or caught into them.

The $E=mc^2$ formula has as much to do with transforms of mass into energy and energy into mass, as does the formula $E=mv^2/2$ for the kinetic energy of a moving body. These two equations express *the same proportionality relations* between mass and energy, and have nothing to do with the unreal mass-energy transforms. Hence, the restoration of physics as a natural science requires the re-establishment of the Law of Mass Conservation and Indestructibility as the most important and unswayable conservation law in science.

The constancy of mass is what makes it differ from other magnitudes, related to the quantity of matter, i. e., from the volume, which changes with temperature and pressure, and from the weight, which changes with location on Earth, distance in space, and buoyancy.

3.06. Do Pan Scales Directly Measure Mass?

Contrary to statements in physics texts, weighing on pan scales does not directly measure the mass. The difference between weighing on spring scales and on pan scales is that pan scales **compare** or balance the weight, put on one pan, with the weight on the other, while spring scales compare the applied force or weight with the elastic restoring force of their spring.

Weighing a body in different locations on Earth on spring scales yields the differing values of weight, depending on the geographic latitude and elevation. Weighing on pan scales yields one value in each of the different locations, because any change in the gravitational force affects in the same way the bodies on both pans. For example, if we balanced the weight of a body on a mountain in the equatorial area by a weight of 1 kg, then on sea-level far north, the increased gravitational force acting on the body, is again balanced by the same weight of 1 kg, which increased identically.

However this invariance of results does not mean that the pan scales directly yield the invariant mass. Firstly, the results are reduced by the buoyant force, the more, the larger the difference in volume between the body and the balancing weight. If pan scales were to measure mass, then the results should be independent of the buoyant force, as is the mass. In our example above, the buoyant forces, which are larger at sea level than on a mountain top, may cause a difference in weight.

Moreover, if pan scales were to measure mass, then they should have been able to do so also in an orbiting satellite. There the masses of bodies remain the same as on Earth. However, the masses cannot be measured there, on spring scales and pan scales alike.

3.07. Unit of Mass

When the metric system was established (by 1800), the unit of mass was a gram (1 g), equal to the mass of a cubic centimeter of pure water at a temperature of 4 degrees Celsius. The official standard of comparison was prepared as a platinum alloy cylinder of volume ~46 cm^3, which was filed and polished until it balanced a liter (1 l = 1000 cm^3) of pure water at 4°C. Hence, the mass of the official standard is 1000 g, or, replacing 1000 by the metric prefix *kilo* (k), it is one kilogram (1 kg).

The kilogram was accepted (in the 1960's) as the basic unit of mass in the International System (SI, *Système International*) of units. It is unfortunate that the name of the unit (and only of this basic unit) contains the multiplying prefix *kilo*. This is a source of confusion for the lawful (or rather "ruleful") use of metric prefixes.

One may ask, e.g., if 1000 kg is one kilokilogram? But the rules are against double prefixes (e.g., no kilokilo, but mega, M). Thus, 1000 kg should be 1 Megagram (1 Mg), but then the basic unit seems to be the g, not the kg. And what is a thousandth of the unit, 0.001 kg? Is it a millikilogram, with the double prefix, or just 1 g. But then the gram again appears as the basic unit.

Another source of confusion is that the kilogram has been used since the French revolution as a unit of weight, i.e., a unit of force. It is not an SI unit, thus not used in physics (not "legally") but useful because known to everyone, like a pound in the US. By definition, a kilogram force (a kgf) is the weight of the kg standard of mass at its location in Sévres by Paris (or on sea level at the 45th parallel).

Nothing is wrong in accepting different features of one standard body as units of different physical quantities. It would also be admissible to establish the volume of the platinum alloy cylinder as a unit of volume, its electrical and thermal resistances as units of these quantities, and so on. The wrong thing would be to use for these units the word kilogram, too, instead of giving each of them a separate name.

With the kilogram, the confusions could be avoided by leaving it as a non-SI unit of force (it anyhow is, and very lively), and by granting the unit of mass a well deserved separate name. For a non-confusing name of the unit of mass we dare to suggest the lav, in honor of Lavoisier. We could then have a kilolav (1 kL) and a millilav (1 mL), together with a clear distinction between mass and weight, which are intrinsically distinct physical magnitudes.

Chapter 4.

DENSITIES OF MATTER

4.01. Density of Matter in Atomic Bodies

The density of matter in a body is defined as the mass per unit volume of the body. If matter is distributed everywhere in the body equally, thus uniformly or homogeneously, the density of matter in the body can be calculated by dividing the mass of the body by its volume. The short notation of this rule is

$$\text{density} = \text{mass} / \text{volume}.$$

The condensed rule, or formula, is obtained by assigning to
 the density of matter in the body - the letter d;
 the mass of the body - the letter m;
 the volume of this body - the letter V.
The formula is then
$$d = m/V.$$

Let us calculate the density of water. The metric units were chosen so that the mass of a cubic centimeter of water would be 1 gram. Hence, the density of water, at 4° Celsius, is 1 gram per cubic centimeter. The mass of 1000 cubic centimeters, or of one cubic decimeter, or one liter of water, is 1000 g, or one kilogram. Hence, the density of water is

1 kg per liter, or $1 \text{ kg/l} = 1 \text{ kg/dm}^3 = 1 \text{ g/cm}^3$.

Let us now calculate the density of matter in the Sevrian standard. We arrange the calculation as follows:

m=1 kg $\qquad$ $d = m/V$
V=46 cm^3 $\quad$ d=1000 g/46 cm^3 = 21.7 g/cm^3 .

The density of matter in the body, composed of the platinum alloy, is also the density of the substance, the alloy, and is 21.7 grams per cubic centimeter, or kilograms per cubic decimeter.

The g/cm^3 and kg/dm^3, or kg per liter, are convenient metric units, but they are not the SI units. The SI units must be composed of the basic units of the system, and not of sub-units. Hence, the SI unit of density is the kilogram per cubic meter. The density of water is then 1000 kg/m³, and the density of the platinum alloy is 21,700 kg/m³.

The use of the SI unit of density, kilogram per cubic meter, leads to large numerical coefficients. Hence, the unit is inconvenient, as are also some other SI units. But it is so in life, that wanting to enjoy the advantages of standardization, unification, and internationalization, we must pay a price here and there.

Note. The density of matter is a distinct physical magnitude. Though determined as mass divided by volume, it is neither mass nor volume.

4.02. Relating to a Physics Formula and Using It

A formula is not omnipotent. It is just a condensed notation of the calculation rule or law, and does not show the restrictions and limits of their applicability. Before using a formula, one should re-read the law and the rule(s) to make sure that they can be applied in the given case.

When substituting the particular values into the formula, one has to make sure that they are the appropriate ones. It quite often happens that in the fever of computing, one just substitutes any handy available data. In the density formula, he would substitute the mass of one body and the volume of another body, e.g., the volume of a liquid in which the body is immersed.

One should be aware that physical values and data (physical magnitudes) consist of a numerical factor (number) and of the appropriate unit(s). The number without the unit has no physical meaning (except in ratios). Thus, the numbers must be substituted into the formula together with the units. Then all mathematical operations, called for by the formula, must be performed on both the number and the unit.

If, e. g., the formula for the area A of a square is $A=l^2$, and $l=13$ m, we substitute for l into the formula 13 m, not just 13; then, because the formula calls to raise l into the second power, we must raise into the second power the number 13 and the unit m as well, to obtain

$$A = (13 \text{ m})^2 = 170 \text{ m}^2.$$

If the units (dimensions) <u>obtained</u> in the result would not have been appropriate (e.g., units of length instead of area), then this would serve as a sign that we did something wrong.

Note that in the result we wrote 170 as the number of square meters, not 169. We did so because the number 13 is a result of a measurement, and is thus an approximate number. According to the mathematical rules, the result of multiplication of approximate numbers must not contain more significant digits than the multiple. The number 13 contains only two significant digits, so that the product, 13·13, is 170, which also contains two significant digits, 1 and 7.

If the number 13 were accurate to 3 significant digits, then this could be pointed out by writing 13.0; here the zero behind the decimal point is understood as a significant digit, too. Then we would have

$$13.0 \cdot 13.0 = 169.$$

Note that every letter in a formula replaces a sentence, identifying a physical magnitude. If the letter is not carefully inscribed, with its right shape and size, with all its appropriate indices, then the physical magnitude may be misidentified, which leads to errors.

4.03. Densities of Substances

The density of matter in a body, composed of one substance, is also the density of matter in this substance, or for short, the density of the substance. Thus, the density of matter in the Sevrian mass standard is also the density of the platinum alloy of which it is made.

Densities of substances are listed in most physics texts and handbooks. They are usually expressed in the convenient gram per cubic centimeter units. The densities of solid atomic substances vary: from ~0.1 g/cm^3 for balsa wood, through 0.53 g/cm^3 for the least dense, or "lightest" metal lithium, to 22.57 g/cm^3, for the densest, or "heaviest" substance on earth, which is the metal osmium.

The metric non-SI units of density are convenient, because the densities of substances have the same numerical values when expressed in g/cm^3, in kg/dm^3 or kg/liter, and in $tons/m^3$. For water, e.g., we have,

$$1 \text{ ton}/\text{m}^3 = 1000 \text{ kg}/1000 \text{ l} = 1 \text{ kg}/\text{l} =$$
$$= 1000 \text{ g}/1000 \text{ cm}^3 = 1 \text{ g}/\text{cm}^3.$$

We shall therefore use these convenient units. However, when substituting into a formula, where SI units only should be used, we must replace the g/cm^3 by $1000 \text{ kg}/\text{m}^3$, because

$$1 \text{ g}/\text{cm}^3 = 10^{-3} \text{ kg}/10^{-6} \text{ m}^3 = 1000 \text{ kg}/\text{m}^3 \ .$$

4.04. The Mass-Volume-Density Connection

The mass of a body, composed of a substance of known density, can be calculated if the volume of the body is given. The mass of the body is then equal to the density of the substance (which is the density of matter in the substance and in the body), multiplied by the volume V of the body,

$$m = d \cdot V.$$

For example, if a body of copper has a volume of 30 cm^3, then with the known density of copper ($d=8.9 \text{ g}/\text{cm}^3$) one can calculate the mass of the body, as follows:

$d=8.9 \text{ g}/\text{cm}^3$ | $m = d \cdot V$.
$V=30 \text{ cm}^3$ | $m=(8.9 \text{ g}/\text{cm}^3) \cdot 30 \text{ cm}^3 = 270 \text{ (g}/\text{cm}^3) \text{ cm}^3 = 270 \text{ g}$

The volume of a body of known substance can be found if the mass of the body is known. To find the volume, one has to divide the mass of the body by the density of the substance,

$$V = m/d \ .$$

For example, if a body of aluminum ($d=2.7 \text{ g}/\text{cm}^3$) balances a 1 kg weight, then its mass is 1 kg; the volume of the body is then:

$d=2.7 \text{ g}/\text{cm}^3$ | $V = m/d$
$m=1 \text{ kg}$ | $V=1000 \text{ g}/(2.7 \text{ g}/\text{cm}^3)=(1000/2.7) \text{ g}/(\text{g}/\text{cm}^3)=370 \text{ cm}^3.$

The most accurate way to determine volumes of bodies and densities of solid and liquid substances is by weighing the bodies in the air and in the liquid(s). The method, using the Archimedes' Principle, is well described in physics texts and is out of our scope here.

4.05. Average Density

If matter is not uniformly distributed in the body but in lumps, as in buildings and cars, then, dividing the mass of the body by its volume, we obtain the average density of matter in the body. The average density may sometimes be of special value; e.g., a submerged submarine continues to rest as long as its average density (not the density of steel or of any other substance in it) is equal to the density of the surrounding water. By pumping-in some outside water (into tanks, thus replacing the air in them) one increases the average density of the submarine, causing the vessel to sink. By pushing water out, one reduces the average density, and the submarine rises to the surface.

A boat, or even a closed car, the average density of which is smaller than that of water, is floating. While the air in the car or boat is being replaced by penetrating water, the average density is increasing, and the bodies sink.

The average density is the one we can calculate for electrons, protons, nuclides, and other particles of nuclear matter, as well as for the Earth, the Sun and the planets. For the bodies of the Solar System, the mass can be estimated from gravitational interactions, and the volumes - from telescopic observations. The volumes of stars can be roughly estimated from their luminosity, colors, and other data.

Masses of nuclear particles are calculated mostly from their electromagnetic and nuclear interactions. The sizes of nuclear particles are obtained from experiments, in which the particles are impinged or "bombarded" by controlled streams of other particles. In these experiments, people (and sophisticated devices) count the numbers of scattered particles in relation to the number of particles which went through without being scattered. Simultaneously, the *scattering angles* are measured, by which the scattered particles were diverted, together with the velocities and energies of the particles. Out of these results, it becomes possible to derive the so called "*cross-section area*" of the scatterers for this type of collisions.

It is assumed that the shape of the particles is spherical (this is our best bet!), so that the cross-section area is the area of a circle. Hence, one can calculate the cross-section diameters and radii, which are then accepted (or not) as the radii of the scatterer particles.

4.06. Average Densities in the Solar System and in Atoms

The average density can also be calculated for the whole solar system, as well as for single atoms. In these systems of bodies, the mass is not distributed over the volume, but 99.9% of it is concentrated in the central body (Sun, atomic nucleus). The combined mass of all other bodies of the solar system: of planet, moons, asteroids, comets, meteors, is 700 times smaller than the mass of the Sun alone.

In the hydrogen atom, which is the lightest, the mass of the orbital electron is almost 2000 times smaller than the mass of the atomic nucleus (the proton). In the heaviest natural atom of uranium, in which there are 92 orbital electrons, and a nucleus of 92 protons and 146 neutrons, the mass of all orbital electrons is 4760 times smaller than the mass of the nucleus.

Obviously then, in the calculation of the average density of mass in these systems, only the mass of the central body counts; the masses of the other bodies are negligible, being smaller than the probable error in estimating the mass of the central body.

4.07. Average Densities of Bodies in the Solar System

Basic data describing the main bodies of the solar system and their average densities are listed in Table 3.

The mass of all planets is $2.7 \cdot 10^{27}$ kg, insignificant in comparison with the 740 times larger mass of the Sun, which is $1.984 \cdot 10^{30}$ kg. Assuming spherical symmetry and accepting for the radius of the solar system the orbital radius of Pluto, $6 \cdot 10^{12}$ m or 6 billion kilometers, we find the volume of the Solar System to be not less than $9 \cdot 10^{38}$ m^3.

Table 3. Solar System Data

	Mass	Diameter	Average density	Orbital radius	Orbital velocity
	in units of				
	10^{24} kg	Mega-m	g/cm^3	Giga-m	km/s
Sun	$2 \cdot 10^6$	1390	1.4		
Mercury	0.31	5.0	4.8	58	48
Venus	4.9	12.4	4.9	108	35
Earth	6.0	12.8	5.5	150	30
Moon	0.074	3.5	3.4	0.38	1
Mars	0.65	6.8	4.0	228	26
Jupiter	1900	143	1.34	778	13
Saturn	570	119	0.71	1430	9.6
Uranus	88	51.5	1.27	2870	6.9
Neptune	103	49.9	1.58	4500	5.4
Pluto	5.4			5900	4.8

The volume of the Solar System is thus six hundred billion times larger than the combined volumes of all bodies of the solar system. The average density of matter in the solar system is then 2.2 10^{-12} g/cm^3. This is 450 times less than the density of water, and comparable to the highly rarefied densities, which can be achieved in the best laboratory vacuum apparatus.

We see that the solar system is space-like empty. The distances between the central body and the orbiting bodies are hundreds of times larger than their diameters. Hence, approximately a trillionth part only of the volume of the solar system is occupied by its bodies. The rest of the volume, i.e., almost all of it, is just as empty as space.

4.08. Average Density in Nuclear Particles

Basic data for some nuclear particles are listed in Table 4.

Table 4. Basic Data for Some Nuclear Particles

	Mass		Radius	Density	Electric charge	
	in	units		of		
	10^{-31}kg	amu	fm 10^{-15}m	10^{14} g/cm^3	10^{-19} Coul	e
Electron	9.1	0.0005	0.09	3	-1.6	-e
Proton	16726	1	1.1	3	+1.6	+e
Neutron	16749	1	1.1	3	0	0
Deuteron	33433	2	1.26	3	+1.6	+e
Alpha-particle	66441	4	1.59	3	+3.2	+2e

<u>Notes</u>. 1. The atomic mass unit (amu) is $1.66 \cdot 10^{-27}$ kg.

2. The new experimental finding of the radius of the electron puts its upper limit at 0.1 **fermi**. A fermi is a femtometer, symbol fm. The value of ~0.09 fm, used here, which is just below the upper limit, yields for the density of matter in the electron the same value of $3 \cdot 10^{14}$ g/cm^3 as for all other listed particles.

The radii R of atomic nuclei are connected with their masses m by the formula

$$R = A^{1/3} \cdot 1.1 \text{ fm};$$

here A is the number of atomic mass units (amu) of the nucleus. Hence, the radii of atomic nuclei vary from 1.1 fm in hydrogen, to ~8 fm in uranium 238.

In the spherical shape approximation, volumes are proportional to R^3. It therefore follows from the formula for the radii of atomic nuclei that the volumes of nuclei are also proportional to their masses. This means that a nucleus of larger mass has a proportionally larger volume. Thus, in all nuclei the ratio of mass to volume is the same. Hence, all known particles of nuclear matter have a similar density of

$$d = 3 \cdot 10^{14} \text{ g/cm}^3 .$$

Chapter 5.

DENSITIES AND STRUCTURE OF MATTER

5.01. The Space-Like Emptiness of Atoms

While the radii of atomic nuclei vary from 1.1 fermi (1.1 fm) to ~8 fm, the radii of the outermost orbits of the electrons in their ground states, i.e, when the atoms are not excited, do not differ much, and are of the order of ~100 picometers (100 pm = 10^{-10} m), which is 1 **angstrom**. The **angstrom**, symbol Å, 1 Å=10^{-10} m, is a useful (but not SI) unit for the atomic size. The angstrom sizes of atoms can now be directly observed and evaluated in the new technique of Scanning Tunneling Microscopy (STM).

The volumes of atoms are ~$4 \cdot 10^{-30}$ m^3, and the average densities of matter in them vary from 0.4 g/cm^3 in the atom of hydrogen, to ~100 g/cm^3 in the atom of uranium 238. The average density of matter in atoms is thus ~10^{14} or a 100 trillion times smaller than in the nuclei and in other particles of nuclear matter.

In atomic solids, even in the "close packed" ones, there is still some space left between the atoms, so that the density of matter in them is a quadrillion (10^{15}) times smaller than in nuclear matter. This means that in atomic solids, only a quadrillionth (a 10^{-15} part) of the volume is occupied by particles of which the atoms consist. The rest, i.e., almost all the volume of atomic bodies is as empty as space.

The emptiness of atoms and of atomic bodies was first discovered by E.Rutherford in his experiments (1911) with the scattering of alpha-particles in metal foils. He found that almost all impinging alpha-particles passed through the foil without changing direction. Only an extremely small number of alpha-particles was scattered, but at unexpectedly large angles. This proved that the atoms in the foil are almost fully penetrable to alpha-particles, i.e., almost empty of bodies that could stop or scatter the alpha particles. It also proved that the whole mass of the atom is concentrated in a nucleus, the diameter of which is a hundred thousand times smaller than the diameter of the atom. Only a collision with such a small and dense nucleus could cause the large-angle scattering.

The 10^{-14} part of the volume of the atom, which is "filled" by the nucleus and electrons, is hundred times smaller than the 10^{-12} part of the volume of the Solar System, which is occupied by the Sun and planets. We thus see, that as space-like empty as the Solar System is, with its bodies so far apart from one another, the atoms are still a hundred times emptier and more space-like.

5.02. The Electron Volt and Other Energy Units

Energies in atomic and nuclear processes are conveniently expressed in electron volt units. The electron volt (symbol eV) is the energy gained (or lost) by a free electron on crossing a voltage of 1 volt (1 V). The standard energy unit in the International System of Units (SI) is the **joule**, symbol J. The joule, or **watt·second**, is equal to the energy used, or work performed, by a device of power equal to 1 watt during 1 second. One joule equals 6.24 quintillion electron·volts. A quintillion is a billion billions, or ten to the 18th power, or the number 1 followed by 18 zeros. Hence,

$$1 \text{ J} = 6.24 \cdot 10^{18} \text{ eV}.$$

Inversely, an electron·volt is 0.16 of a quintillionth of a joule. A quintillionth is ten to the minus 18th power, or the number 1, in front of which there are 18 zeros, including the zero before the decimal point.

Thus, 1 eV = $0.16 \cdot 10^{-18}$ joule, or better yet,

$$1 \text{ eV} = 1.6 \cdot 10^{-19} \text{ J}.$$

To a diet-minded person, the familiar energy unit is the calory. The dietician's **calory** equals 4180 joules. Hence in a calory, there are $2.6 \cdot 10^{22}$ electron volts, and 1 eV is $3.83 \cdot 10^{-23}$ of a calory.

If one does not watch calories, but pays electric bills, he is familiar with another unit of energy, the **kilowatthour** (symbol **kWh**), for which he pays about 10 cents. The kilowatt·hour is the energy used (or work performed) by a device of 1 kilowatt power during 1 hour. Thus, 1 kWh = 3,600,000 watt·seconds or joules, or

$$1 \text{ kWh} = 860 \text{ calories} = 2.24 \cdot 10^{25} \text{ eV, and}$$

$$1 \text{ eV} = 4.5 \cdot 10^{-26} \text{ kWh}.$$

5.03. Energies of Orbital Electrons and Atoms

The binding energy of the electron in the hydrogen atom in the normal or "ground state" is 13.6 electron·volts. In all other atoms, the "ground state" energy of the outermost orbital electron is of the same order of 10 eV. The energy of the electrons on inner orbits is the higher the closer the orbit to the nucleus, and the larger the number of protons in the nucleus, i.e., the larger the "atomic Z-number". For example, uranium atoms are the heaviest natural atoms, their Z-number is 92, and the binding energies of their innermost orbital electrons are thousands of electron volts.

In a cubic centimeter of a solid there are $\sim 2 \cdot 10^{22}$ (20 billion trillions) atoms. Hence, the energies of only the outermost orbital electrons in all these atoms is well over 10^{23} eV, or $\sim 30{,}000$ joules, or ~ 8 calories. This is many times more than the energy needed to melt, boil, and evaporate a cubic centimeter of any known solid.

The average energy of an atom (or simple molecule) in its thermal motion at room temperature in a solid, liquid, or gas (*"mean thermal energy"*) is several hundredths of an eV (e.g., 1/40 of an eV, or 25 milli-eV). This energy is hundreds of times smaller than the energy of only the outermost orbital electron of the atom.

Nevertheless, the mean thermal energy of an atom is still many times larger than the average energy which we submit to an atom of a body, when we compress or stress the body in our strongest mechanical presses or pumps. All such compressions or stresses do not affect the size of atoms but reduce or increase the inter-atomic distances only.

5.04. Why Are the Empty Atoms Impenetrable To One Another?

When we press our finger against any atomic body, soft or hard, we act with per-atom energies which are very far from affecting the outer electron orbits of the atoms in the body and in our finger. If the body is soft, then its grains, cells, molecules, and atoms move apart, giving in to the pressing cells of the epidermis of the finger, with the atoms and molecules unaffected. Even the cells remain intact, or else it hurts! If the body is hard, then the cells, with their constituent molecules and atoms, move slightly apart, giving in to the pressure. Slightly, of course, else we get bruises and cuts. In any case, the molecules and atoms remain unaffected, with no penetration whatsoever of atoms into one another.

Atoms cannot penetrate into one another, not even slightly, because the energies of their mechanical and thermal motions are much smaller than the binding energies of their outermost orbital electrons. In addition, the outermost orbiting electron creates a shell of negative electric charge around its atom. When two atoms approach one another too closely, then the energy of the electric repulsion between their outer shells of negative charge will eventually become larger than the energy of their thermal and mechanical motion or pressure.

Another view is obtained by considering that the outer orbital electron in non-excited atoms moves with a velocity of two thousand kilometers per second, and performs quadrillions of rotations per second. The plane of the orbit is not stationary in space but vibrates and rotates around its axes, though with million times lower frequencies. The plane of the orbit performs billions of full rotations per second. Hence, during a billionth of a second (a nanosecond), the orbital electron sweeps-over the whole closed outer surface of the atom. In other words, it takes the orbital electron a nanosecond to "visit" every single point on the outer surface of the atom.

If two atoms approach each other and "touch" during a nanosecond, then their outer electrons succeed to cover-up the whole outer surfaces of their atoms. Each electron creates a sphere-like electrically charged cloud which is impenetrable to the cloud created by the electron of the other atom. Even if the outermost electron of one atom succeeds to penetrate the cloud of the other atom, it will be stopped and repelled on the much denser cloud, created in the atom by electrons on inner orbits. This is why the space-like empty atoms are not penetrable to other atoms.

Atoms are also not penetrable to very slow electrons (e.g., to 'free' or conduction electrons in metals and semiconductors), and to some other particles of nuclear matter. For these particles, the orbital electron of the approached atom is always in their way, blocking and repelling. However, fast particles manage to cross the orbital cloud at moments when the orbital electron is not in their way. They do not encounter a cloud but a cross-section area for their scattering by the orbital electron. This cross-section area is the smaller the larger the velocity of the penetrating particles.

Atoms **are** penetrable to protons and alpha-particles; due to the positive electric charge of these particles, they are not stopped by the negatively charged clouds created by the orbital electrons of the atoms. And, due to their large masses, the scattering of protons and alpha-particles on orbital electrons is negligible. When the protons and alpha-particles approach the vicinity of the nucleus, they are repelled by its positive charge. The large-angle scattering occurs on the very rare occasion of head-on collision with the nucleus.

Electrically neutral particles of large mass, like neutrons, can easily penetrate through the whole atom. They are not repelled by the nucleus but usually slide around it. A neutron may also be caught by the nucleus, turning it into an *isotope* nucleus, or initiating its fission.

5.05. Can The So Empty Atoms Be Shrunk?

The distances between the orbital electrons and the nucleus in the atom are so large that to a person of fine general education it seems natural that they could be shortened a hundred times. The distances would then still be a thousand times larger than the diameters of the particles, but the volume of the atom would decrease a million times. Imagine then the great savings in storage capacities and in housing, which can be achieved by shrinking industrial products, cars, people, "as seen on TV" and in the movie "Honey, I Shrunk the Kids".

It is possible that sometime physicists may find ways by which in specially built reactors the orbital radii in some atoms would be somehow reduced. Hence, the straight answer to the problem in the title is "yes, some reduction might become possible". However the energy investment would have to be tremendous.

Suppose that we want to shrink, and only tenfold, the atom of hydrogen, which is the simplest, consisting of one proton and one orbital electron. The binding energy of the electron in the normal or "ground state" is 13.6 electron·volts. To reduce the radius tenfold, resulting in a thousand-fold reduction in volume, one would have to increase ten times the energy of the electron in the atom, by pressing into the atom an energy of 140 electron·volts.

In a cubic centimeter of a solid there are $2 \cdot 10^{22}$ atoms. Even if the shrinkage of these atoms would require the same energy as the shrinking of a hydrogen atom, with its only electron, we would have to press into the cubic centimeter of the solid the energy of 140 eV multiplied by the number of atoms, i.e., $3 \cdot 10^{24}$ electronvolts. This is above hundred calories (of the dietician).

One such calory, submitted to a cubic centimeter of any solid would evaporate it completely, and heat the vapors to a burning high temperature. But the body with the shrunken atoms would be holding an excess of **100** calories per cubic centimeter! Exposing this body to the open air would cause it to explode while evaporating.

Thus, the shrunken atoms would have to remain in the reactor, or in similarly expensive storage tanks. In the free outside world they would explode, releasing the tremendous amount of energy invested into the shrinking. This per-atom energy is several times larger than in the strongest "conventional" explosives.

The binding energies of atoms in molecules are smaller than the binding energy of the outermost electrons in the atoms. The energy pressed into the shrunken atoms is thus much larger than the binding energies of atoms in molecules. Hence, a shrunken atom would tear off even the strongest molecular bonds in any molecule. In other words, shrunken atoms would certainly not be able to form molecules and substances at all, let alone the ones formed by regular atoms and needed to support our lives.

We remark that the usual compressors reduce the interatomic distances, not the atomic radii. The reduction of the volume of a compressed atomic body does not reflect a reduction in the volumes of the atoms, which is **not** achievable.

5.06. Can Atoms of our Bodies be Shrunk or Blown?

Our lives can be preserved when our body temperature is in the narrow range between 20 and 45 degrees Celsius. This corresponds to a maximum increase of energy per cubic centimeter of our body by up to 0.008 calories, and to a decrease not exceeding 0.015 calories. But the tenfold shrinking of atomic radii requires 100 calories per cubic centimeter of the body. This is ten thousand times larger than the killing energy. So please don't play with atomic orbits.

Suppose that we add to each cubic centimeter of the body slightly less energy than the killing 0.008 calories. Even if we found a way to force the atomic orbits to accept this or any other arbitrary amount of energy (they won't!), we would reduce the radius of the outermost orbit of each atom by 0.1%. The volume of the atom might then be reduced by 0.3%. Hence, normal dieting is much more effective also for savings in storage and enclosure capacities for our bodies than trials to shrink atoms, and much less dangerous, too.

Apart from all bans, mentioned and unmentioned, the hundred-fold reduction of sizes in the movie "Honey, I Shrunk the Kids" would require thousand calories per cubic centimeter, or 50 million calories per kid, or 60,000 kilowatthours, at a cost of 6000 dollars in electric bills. But the first 30 thousand calories, or 40 kilowatthours altogether, for the cost of 4 dollars, would have completely evaporated the body. Hence, the movie suggests a way of great savings in burial costs.

We make this macabre point to accentate the absurdity of the "science" fiction, which is constantly emerging in movies, TV programs, newspaper "science" reports, and books. They have usually nothing to do with real science. At best, they may be based on mathematical misinterpretations of experimental results. A basic knowledge of natural physics would enable us to reject them instead of letting them affect our thinking, and budget.

The budget, and the income of the producers of such fiction is tremendous, on society's account (*"there are no free lunches!"*; and if somebody gets it, then some others somehow pay for it). And after each pseudo-science nonsense movie or book, there come new ones, in series. So after the shrinking of kids came the movie blowing them up, in all directions. Wait and see movies expanding kids in one or two (or six?) dimensions. Don't we need some real natural physics in our educational system?

5.07. Aggregation States of Atomic Matter

Atomic matter exists in three different **states** of aggregation: the gaseous, liquid, and solid state. Within a state, some substances may appear in different **phases**, depending on the conditions. Such are the superfluid phase, the amorphous, crystalline, superconducting phase, various magnetic phases, etc.

In the normal gaseous state, the molecules (or single atoms in *monatomic* gases) are not bound to one another. The molecules and atoms move freely and independently, except during collisions. If even at such moments of intimate closeness the molecules or atoms do not stick to one another, then the collisions are elastic, and the gas is **ideal**.

In a cubic centimeter of an ideal gas, (at so called *normal conditions*, i.e., temperature of zero degrees Celsius and normal atmospheric pressure), there are 27 quintillion ($2.7 \cdot 10^{19}$) molecules. The mean distance between molecules in the gas is 33 angstroms. This distance is ~30 times larger than the diameter of the atoms, and 10 to 15 times larger than the diameters of molecules. Therefore the MEAN FREE PATH, traveled by a molecule in the gas between two head-on collisions, can be tens and hundreds of times larger than the mean distance between molecules.

When a gas is below its *critical* conditions of temperature and pressure, it **condenses** into a liquid or, in some substances, into a solid. In liquids and solids, collectively named "**condensed matter**", the distances between the <u>centers</u> of the molecules, or atoms, or ions, of which the bodies consist, is about 3 angstroms, (~3 Å), 10 times shorter than in gases, and the volume occupied by any number of molecules is ~1000 times smaller.

Thus, in a cubic centimeter of "condensed matter" there are $\sim 2 \cdot 10^{22}$ molecules, atoms, or ions. The smallest visible particle, of size 20 microns, has a volume of

$$\sim 3 \cdot 10^4 \ \mu m^3 = 3 \cdot 10^{-9} \ cm^3 \ .$$

The number of atoms or molecules in it is thus
$$3 \cdot 10^{-9} \text{ cm}^3 \cdot 2 \cdot 10^{22} \text{ molecules/cm}^3 = 6 \cdot 10^{13} \text{ molecules},$$
or 60 trillions (millions of millions) of them!

In gases, the molecules do not stick to one another and can cross distances of hundreds of angstroms between collisions with other molecules. However in liquids, and especially in solids, the molecules, atoms, or ions are strongly bound to each other and can only <u>vibrate</u> around certain equilibrium sites, like the "lattice sites" in crystals.

The average (thermal) velocities, with which molecules cross these sites in a solid, are comparable to the velocities of <u>such</u> molecules in a gas at equal temperature. However, being bound, the particles of the solid must remain within the 3 angstrom distances between their centers. Hence, in condensed matter the constituents have no "free path", in spite of their large thermal velocities. The maximal displacements (amplitudes) in their vibrations are shorter than the 3 angstrom distance between the centers of the constituents. They may move farther away from the equilibrium position or lattice site, only after gaining an energy larger than their binding energy.

5.08. Does Aggregation State Affect Chemical Composition?

The physical properties of a substance change significantly when the aggregation state changes. Some properties become extremely different, quite often also in different phases of the same state. To a solid-state physicist, amorphous and single crystalline silicon, both solids, are completely different materials, though chemically they are considered as one substance.

So are also carbon black, graphite and diamond, all solid, built of carbon atoms only. However to a physicist, they represent completely different materials, and not only to physicists. To glaziers, jewelers, and lovers, diamonds are best friends. All those people would hardly believe that what they work with, sell, pay for, and wear, is one substance with carbon black or graphite.

The assumed chemical unity of a substance in different states of aggregation can also be questioned, and in many cases this unity assumption was proven wrong. Substances consisting of diatomic (two-atom) molecules in the gaseous state, can have in the solid state molecules of 6, 15, and 30 atoms (e.g., sulphur), or may turn monatomic, even ionic. Hence, by solidification or evaporation, they turn into different chemical compounds.

Monatomic gaseous metals, like copper (Cu), solidify into a piece of metal, representing one giant molecule. In it, the outer orbital electrons of each atom are 'freed' and belong collectively to the remainder positive ions of the metal, e.g., to all the Cu^+ ions in the piece of copper.

Certain texts classify phase transitions within a given state, and the changes of aggregation states, i.e., solidification and melting, sublimation, evaporation and condensation, as physical processes, together with dissolution, heating and cooling, compressing, magnetizing, electrifying, etc. The texts also define physical processes as such in which the chemical composition of the involved bodies does not change, i.e., that the chemical substance(s) of which each body consists remain(s) the same.

However the much more sensitive new chemical methods and sensors reveal the changes in chemical structure occurring in many processes, which were considered purely physical. Some chemical reactions occur in a certain state of aggregation, and do not occur in the other states. Hence, chemical compositions do not always remain intact in physical processes. This, too, points towards the chemical non-unity or diversity of the substance in the different states and shows that there are no sharp border lines between physical and chemical processes and phenomena.

5.09. Motion in Fluids

When liquids and gases, collectively named "fluids", move in other fluids or mix with them, whether slowly or in fast jets, their atoms or molecules make their way <u>between</u> each other, through the interatomic or intermolecular spaces. The atoms and molecules do not penetrate through each other. If they interact chemically, then they may exchange or bind their outer orbital electrons (valence electrons), without penetrating any further through or into each other.

Just as atoms are not penetrable to other atoms, solids are not penetrable to other solids and to fluids. The binding between the constituent atoms, ions and molecules in the solid is so tough that even the interatomic and intermolecular spaces in a solid are not penetrable to atoms of the other solid. A certain penetration, to limited depths, occurs at high temperatures. Light atoms, such as lithium, and especially hydrogen, penetrate easier, but also not throughout a thick solid.

Thus, when a solid atomic body moves through a fluid (i.e., through a liquid or gas) even with a jet-plane or missile velocity, the atoms or molecules of the fluid cannot penetrate through the solid. To move, the solid must push apart entire layers of the fluid on its way, inevitably causing currents or winds in the fluid. The motion faces a resistance, which is larger the higher the velocity of the body.

5.10. Motion of Atomic Bodies Through Collectives of Nuclear Particles

Bodies of atomic matter can move through a collective of nuclear particles without causing winds in it. The distance between atomic nuclei in a solid is ~3 angstrom, a hundred thousand times larger than the diameter of the proton or neutron, and 20,000 times the diameter of the largest stable nuclear particle known on earth - the uranium 238 nucleus, which is ~8 fermi (~8 fm).

A very thin metal foil may therefore be compared to a net, the eyelets of which are a hundred thousand times wider than the cords, and twenty thousand times larger than the "catch", or the particles of the ambient. Therefore, the particles of the ambient cannot be "caught" by the net. The motion of the net in this ambient would not face resistance and would not make winds or currents in the ambient.

Let us consider a "**reversed Rutherford experiment**", i.e., not with a stream of alpha-particles impinging on a metal foil, but with a metal foil moving through a collective of alpha-particles. Almost none of these particles, met by the moving foil, would be disturbed by it. The motion would be resisted only by those alpha-particles which would be on the way of the atomic nuclei proper. But the distances between the nuclei in the foil are 50 thousand times larger than the sizes of the alpha particles. Hence, the motion of the foil would not cause noticeable winds in the collective of the alpha-particles.

The motion of a foil through a collective of neutrons would proceed with even less resistance, because the atomic nuclei would not scatter the neutrons. Thus, not only a foil but also a thicker solid body could move through a collective of neutrons without making winds in it.

The smallest thinkable resistance to the motion of a solid body would be exerted by a collective of electron positron pairs, such that in each pair the interparticle distance is ~5 fermi (5 fm), which is the size of "medium to large" nuclei. The nuclear particles of the atoms of the body, which moves in this collective, would "see" each electron-positron pair of the lattice as a neutral particle of very small mass, like a certain *neutrino*. On the other hand, the 5 fm distance between the neighboring electron and positron in such a collective is sufficiently large to avoid head on collisions.

If the particles of the collective form the electron positron lattice (epola), then the lattice bonds between particles in equilibrium are very "soft", not rigid. Thus, an atomic electron or nucleus, even the one moving on a "collision path" with a particle of the lattice, can easily make its way between the bound particles of the lattice, pushing them slightly apart. Hence, motion of atomic bodies through a collective of such electron-positron pairs, forming the epola, will cause waves in it (de Broglie "waves of matter", introduced by L.V.de Broglie in 1924) but cannot cause any winds in this ambient.

5.11. Are there Aggregation States of Nuclear Matter?

The density is the characteristic physical property of atomic matter which undergoes significant, even drastic changes, when the aggregation state changes. The density is also a measurable characteristic in nuclear matter. However all stable particles of nuclear matter, known on earth, from the electron, and up to the 440,000 times more massive nucleus of uranium 238, have a similar density, close to $3 \cdot 10^{14}$ g/cm^3. The unstable particles, the lifetimes of which are sufficient to allow measurements of mass and cross section areas, also seem to have this density value. Assuming that the density should change with change of state, we might suggest that

all stable and quasi stable nuclear particles, known on earth,
belong to a single aggregation state of nuclear matter.

Matter in nuclear stars seems to have a not very different density. A neutron star, observed and studied since 1987, was found to have a radius of ~10 km and mass ~1.4 times the mass of our Sun. Thus the calculated average density of matter in this star is $7 \cdot 10^{14}$ g/cm^3. The calculated density of the inner core of the star seems to be a hundred times higher, which might point out that the core is in a still denser aggregation state.

The dark matter in space, the mass of which is now calculated by some to constitute 99% of the mass of the universe, the gray matter, and the black holes of different sorts, may represent either **kinds** of matter, distinct from the atomic and nuclear kinds or, possibly, different aggregation states of nuclear matter.

We prove that the "vacuum space" or "empty space", which actually means "space evacuated or emptied of atomic matter" is not at all empty of **any** matter but contains bound particles of nuclear matter. Hence, **our** vacuum space is a kind of matter, different from atomic matter, and from "solid" nuclear matter, as well as from the known unbounded or free particles of nuclear matter. Hence, *the vacuum space can be considered an aggregation state of nuclear matter.*

Let us recall, that atomic matter, including ourselves, consists solely of particles of nuclear matter, as does every single atom. Therefore,
*atomic matter can also be considered
a certain aggregation state of nuclear matter.*

5.12. Improper Generalization of Results on Any Matter

With our bare senses we do not perceive particles of nuclear matter. Though our first encounter with some of them started with the discovery of the radioactive decay of uranium in 1896, it took decades before particles of nuclear matter were singled out, recognized and studied as distinct entities. Still, there is no full recognition of nuclear matter as a distinct kind of matter, the laws of which do not automatically apply to atomic matter, and *vice versa*.

For example, in 1905 Albert Einstein derived the dependence of the mass of an electron on its velocity. He then automatically generalized his result on *"ponderable material points"* i.e., on appropriately small bodies of atomic matter. He "proved" it by saying: *"because a ponderable material point can be made into an electron..."*.

Well, it cannot. No piece of atomic matter can be made into an electron! Einstein's statement wasn't very good even in 1905, when atoms were considered droplets of a positively charged liquid, with negatively charged electrons floating in it. Later, when the tremendous density of matter, and the tremendous density of electric charge in the electron became known, this statement turned out to be an obvious nonsense. Nevertheless, the statement was not corrected until now, even in the most serious physics books.

Moreover, numerous scientific results, found for electrons or other nuclear particles, and good for them, as, e.g., de Broglie waves, are automatically generalized on all matter, on the whole universe, and beyond, with no good reason, but for the "Unity of Matter".

Physics texts, pop science books, and teachers, seem to be more fascinated by the generalized results than by the original findings. They calculate, e.g., the unmeasurably small and meaningless values of the Einstein increase of mass and the de Broglie wavelength for fast boats, cars, planes, without pointing out that these effects were derived for and relate to *electrons* (and other nuclear particles) moving in the electromagnetic field, and not to atomic bodies.

5.13. Deceiving Perceptions and Concepts of Matter

With our bare senses, we do perceive atomic bodies, but not the molecules and atoms of which they are built, and certainly not the nuclear particles of which the atoms consist. When we touch or push a solid body, our fingers cannot penetrate it, because the atoms in our fingers cannot move through the atoms of the body or through its interatomic spaces (thank goodness!). Therefore we perceive atomic bodies as being dense and continuous.

These "natural" perceptions led to the development of our similarly "natural" concepts that all matter is dense and continuous, even gases. Whole branches of mathematics were developed to treat this *"Physics of Continuous Media"*. These concepts and treatments serve us well in everyday life on macroscopic scales. However on atomic and sub-atomic scales they were experimentally disproved by Rutherford in 1911. Atomic matter is not dense, but almost as empty as the vacuum space is considered to be. Atomic matter is not continuous but discrete, and very much so, because it is built of discrete molecules, ions, and atoms, which consist of discrete nuclear particles, far apart from one another.

Unfortunately, these irrefutable facts, known since 1911, did not become part of our picture of the world, nor did they influence our thinking; not of the general public, not of mathematicians in physics, and not even of pure physicists.

When calculations based on the dense and continuous media concepts fail, imaginary dimensions are added, along with imaginary *Riemannian Surfaces*, imaginary *Minkowski Spaces*, with imaginary exotic particles, and what not, just to save the wrong concepts. Sometimes this arsenal of mathematical inventions does yield satisfactory calculative results. But the calculative achievements last usually till a new experimental finding makes them inapplicable. Then new mathematical inventions are applied, instead of turning to look for a right physical model.

The situation is similar to what it was for 1400 years, when the Ptolemaic model was reigning in science. The model was based on our very natural perception that all celestial bodies are rotating around us, as is seen with our very eyes. This perception led to the natural concept that we on our Earth are the center of the universe. The concept of "us in the center of the world" corresponded to the philosophical, religious, and ideological beliefs of those times, so that the geocentric model got full official support.

To cope with this model, mathematical treatments were developed, in which planets had to rotate on hypocycles, the centers of which rotate on epicycles, the centers of which rotate on orbits around us. If the calculative result did not fit, one could always add as many rotary motions and cycles as needed for a fit. So this is the way it **was** and **is**, because this is the way **we are**.

The belief in "us in the center of the world" undergoes a revival in the now ruling astrophysical theories. The theories maintain that the universe was created in a **Big Bang** explosion at a point near to where we are now, and that the poor thing continues to explode. Hence all things in the universe, galaxies etc., run away from us, while we remain in a good safe place by the center.

tions and atoms which consist of electric nuclear particles, far apart from one another.

Unfortunately, these irrefutable facts, known since 1911, did not become part of our picture of the world and they influence our thinking, not in the general public, not of mathematical nature, and not even of purely scientists.

When calculations based on the dense and voluminous media concept fail, imaginary dimensions are added, along with imaginary Riemannian Surfaces, imaginary Minkowski Spaces, with imaginary exotic particles, and when not, just to save the wrong concept. Sometimes this arsenal of mathematical inventions does yield satisfactory calculative results, but the calculative achievement has usually off a new experimental finding makes them inapplicable. Then new mathematical inventions are applied, instead of daring to look for a right physical model.

The situation is similar to what it was for 1400 years, when the Ptolemaic model was reigning in science. The model was based on our visual, natural perception that all celestial bodies are rotating around us, as is seen with our own eyes. This perception led to the natural concept that we on our Earth are the center of the universe. The concept of us as the center of the world corresponded to the philosophical, religious, and ideological belief is of those times, so that the geocentric model got full cultural support.

To cope with this model, mathematical inventions were developed, in which planets had to rotate on hypocycloidal centers of which rotate on epicycloids, the centers of which again in subtle rounds. If the calculative result did not fit, one could always add as many rotary motions and cycles as needed for a fit. So this is the way it was and is because things, the way we are.

The Bauch in "us in the center of the world" underwent a revival in the now ruling astrophysical theories. The theories maintain that the universe was created in a Big Bang explosion at a point near to where we are now, and that the good thing continues to explode. Hence all things in the universe, galaxies, etc., run away from us, while we remain in a good safe place at the center.

CIRCLE TWO

Chapter 6
ABOUT PHYSICS AND OTHER NATURAL SCIENCES

6.01. What is Physics?

The word physics originates from the Greek *physis*, meaning NATURE. Hence, as a science, physics is meant to deal with <u>bodies</u> existing in nature, or NATURAL BODIES, and with <u>phenomena</u> occurring to them, or NATURAL PHENOMENA.

Sciences dealing with natural bodies and phenomena are NATURAL SCIENCES. These are, in addition to physics, also chemistry, biology, astronomy, and geology. Each natural science deals with certain groups of natural phenomena, assigned to this science. For example, physics deals with PHYSICAL PHENOMENA, chemistry - with chemical phenomena, and biology - with biological or life phenomena.

Similarly, each natural science, except physics, is concerned with a certain group of natural bodies (MATERIAL bodies) assigned to this science. Astronomy deals with astronomical (celestial, or cosmic) bodies, biology deals with live bodies, chemistry deals with chemical bodies or substances. Physics is concerned, to a certain extent, with ALL natural bodies, small and large, dead or alive.

Therefore, all material objects are **physical bodies**, from NUCLEAR PARTICLES to the million times larger ATOMS, IONS, and MOLECULES; from viruses and bacteria, to our own million times longer bodies; from the smallest specks of dust, to our million times larger cars and houses; from our cars and houses, to the million times wider seas and continents; from the Earth and planets, to the million times larger stars; from stars to the million times larger galaxies; from galaxies to their million times larger clusters, and from them to whatever else is

49

there, in whatever is the physical body named Universe.

6.02. Physical Phenomena

There are five main groups of physical phenomena, i.e., of phenomena assigned to and dealt with by physics. These are:

MECHANICAL PHENOMENA, like inertia and gravitation; friction, buoyancy, and deformations; mechanical motions: recti-linear, circular, rotational, vibrational, undular (wave motions).

THERMAL PHENOMENA, like heating and cooling (heat transfer); thermal motion of molecules, atoms and ions, in gases, liquids, and solids; evaporation and condensation; melting, solidifying, and crystallization.

ELECTRIC PHENOMENA, like charging and discharging of bodies; ionization of atoms and molecules; attraction and repulsion of charged bodies, as electrons, positrons, protons, ions, and other CHARGE CARRIERS; motion of charge carriers in electric fields, and electric currents.

MAGNETIC PHENOMENA, like magnetizing and demagnetizing of bodies; attraction and repulsion of magnetic bodies and of electric currents.

RADIATION PHENOMENA, including emission, propagation, absorption of: sound, infrasounds, and ultrasounds in atomic matter;
electromagnetic radiation: radio- and microwaves, optical (infrared, visible, ultraviolet), X- and gamma rays;
particle beams: of nuclear particles, and of atoms, ions, molecules.

In addition, there are numerous COMBINED PHENOMENA: mechano-thermal, thermo-mechanical, electro-mechanical, thermo-electric, electro-thermal, electro-optic, opto-electric, and other.

6.03. Basic and Complex Phenomena

Mechanical motions can be considered the simplest and most basic physical phenomena. Thermal phenomena are based on (or consist of) mechanical motions of the billions of quintillions of atoms and molecules in the bodies. Electric currents and particle beams are based on the mechanical motions of charged and neutral particles. Sounds and electromagnetic radiations are based on vibrations and waves in atomic matter and in space.

The physical phenomena based on mechanical motions should by no means be considered simple sums of these motions. The phenomena are much more complex and cannot be synthesized of the constituent mechanical motions alone.

Physical phenomena are basic to the phenomena dealt with by other sciences. Astronomical, chemical, geological, and biological or life phenomena consist of physical phenomena, are triggered or stopped by physical phenomena, and may also cause them.

Phenomena dealt with by different natural sciences are thus interrelated, mutually dependent, and interwoven with one another. Therefore it is hard, and sometimes impossible to draw exact border lines, separating the domains or phenomena related to one science from those of another science. Thus, "in between" sciences developed, like physical chemistry, chemical physics, biochemistry, biophysics, and astrophysics.

Biological, biochemical, biophysical, even the simpler chemical phenomena are not just sums of the constituent physical phenomena, and cannot be synthesized of them. Chemical, biochemical, and especially biological phenomena are much more complicated than each and all the physical phenomena of which they consist. By analogy, the ability of a machine or person to reproduce all sounds of a language is not equivalent to the ability to talk the language. Talking is not just a sum of the sounds of a language.

Therefore, according to the complexity of the phenomena, with which the main natural sciences deal, biology is the most complex natural science, chemistry is less complex, and *physics is the least complex, the simplest and most basic natural science.*

6.04. Penetration of Physics by Mathematics

Because of the simplicity of physical phenomena, they are the easiest and best subject for the applications of mathematics, which is not a natural science. While mathematics is very helpful in all human activities of modern times, the tools of mathematics are not sufficiently developed yet to present, let alone to solve, major problems in social or life sciences. Mathematics has already become a major problem solver in chemistry, but not to such an extent as in physics.

Mathematics penetrated so deeply into physics that it is often hard to say where one ends and the other starts. On the one hand, this interwovenness is a blessing to the development of both sciences. On the other hand, it resulted in an excessive mathematization of physics.

To most of us mortals, mathematics is incomprehensible, let alone the highly advanced one, employed in new physics. Thus, physics became the most complex natural science, and is avoided by prospective students and readers, who are generally interested in natural sciences. It is our hope that this book may improve the situation.

6.05. Interwovenness of Natural Phenomena

Any event occurring in nature involves numerous phenomena, which are interwoven with one another. This includes not only physical phenomena, but also chemical, biological, and other phenomena. One of the primary tasks of a science is to analyze the observed events and "dismantle" or dissect them into the composing phenomena and interactions. Without the ability to perform such analyses, no serious scientific results can be obtained. Still then, life can go on, and some technological advancement in utilizing the events may continue.

As an example of the interwovenness of phenomena in natural events, let us consider the throwing of a stone (or of a more sportive object: ball, discus, etc.). During the process of throwing, with the stone accelerated by the palm of the hand, there are various mechanical motions, deformations, and heating, of the stone, the hand, and of other parts of the thrower's body.

These motions result from numerous biological, biochemical, and neurological phenomena, with scores of electric currents flowing in the brain and nervous system of the thrower. There may also be social, psychological, economical, political, or racial phenomena involved, which provide the reasons and stimuli for the throwing.

Account of the involved phenomena is helpful in improving the performance and achievements in all sports based on throwing balls, hammers, spears, etc. Detailed analysis and deep knowledge of the motion of thrown bodies is necessary in artillery, rocketry, and in the launching of space vehicles.

6.06. Simplicity and Complexity of Physics

Physics should be the simplest natural science not only because it deals with the simplest natural phenomena, but also because of the way physics deals with natural bodies. In most cases, physics has no concern about the internal structure, chemical composition, and chemical or life processes in bodies. Quite often, physics regards bodies as *material points*, i.e., ignoring even their shapes and sizes, and relating only to their mass.

However in reality, physics is the most complicated and the least understood science. It is also most unpopular and hardest to study. High-school students avoid physics, whenever possible. College students seldom choose physics, though acceptance is easiest and competition minimal. This can be observed even at times when employment prospects are bright, and when being a physicist is honorable.

But if the phenomena and the handling of bodies in physics are the simplest, then why is physics so complicated? This question is similar to the "if it is so good, then why is it so bad?" questions, which people ask under suppressive regimes, even at the risk of prosecution.

Under the "regime" of physics, the answer is, that the chairmen, publishers, authors, teachers, treat the experimental, and especially the mathematical TOOLS of physics as if these tools were the most important, or even the whole content of this science. By enforcing such contents, the "regime" caused physics to be almost as technical as technologies, and often more mathematical than mathematics itself.

The physics literature is now filled with detailed technical descriptions of too many experimental devices and setups, and of endless multipage mathematical derivations and "proofs" of everything. The technical descriptions are hard to overcome to people who lack technical ingenuity, while the derivations and "proofs" are insurmountable to the great majority of us, who do not have the very special gift to think and handle mathematics.

These contents do not belong to the natural science of physics but to the respective technical and mathematical disciplines. Their "invasion" causes physics to be so complicated to "regular" people, who are interested in this natural science and able to handle it <u>as</u> a natural science. Without the burden of such contents, physics is really simpler than any other natural science.

6.07. Effects of Extraneous Requirements on Physics

The requirement to be technical as an engineer and mathematical as a mathematician (or more!) keeps people away from physics, also people who have a keen interest in natural sciences. The absence of such requirements in biology and in chemistry, makes these more complex sciences easier to study, thus more interesting and popular.

As any science (or, for this matter, any enterprise in modern society), physics cannot be serious or successful without making full use of the best in technology and mathematics. But this does not mean that a physicist must be a specialist in these disciplines.

Imagine, how many good people would give up the career of a TV interviewer, if required to be able to operate the TV machinery, and to run the accounting of the broadcasting company. Better yet, how many good people would give up politics, if required to repair voting machines, office computers, or air-conditioners, and to calculate the effects of any taxation scheme on their country, on the needy world, and on the whole universe.

As a "natural scientist", the physicist should be able to study physical bodies, interactions, and phenomena, as they are in nature. He should analyze the observed events, and single out all participating bodies, entities, interactions, and phenomena. He must not neglect any single participant, no matter how small its role. When conditions change, the neglected or overlooked participant may become dominant.

A physicist should seek physical explanations and understanding of the observed phenomena and events, and not be satisfied with just the knowledge of how to handle and how to calculate them. Such knowledge alone is sufficient to appreciate engineers and mathematicians working in physics, but not the physicist.

The physicist, too, should know his technical, mathematical, and other tools, but only insofar as is needed to assess what they can do for his task, and when and how to apply them under given conditions. Anything more should be turned over to engineers, mathematicians, and other specialists.

Excessive mathematical requirements close the way for people, talented in physics, and opens the way to mathematicians. While real physicists work in their laboratories, some mathematicians take over administrative and teaching positions, making way for more mathematicians. These introduce more extravagant mathematical requirements, which eliminate real physicists and promote more mathematicians. Such is the sad story of how the natural science of physics was turned into an applied mathematics.

6.08. The Principles of Natural Physics

The turning of physics into a mathematical science is an irreversible process, impossible to stop. The only way to restore the natural science of physics, seen to this author after twenty years of Don Quixote'ing, is by initiating a separate science of natural physics.

Natural physics is meant to absorb physics, with the **use** of technology and mathematics, but only as needed technical and calculative tools, not as dominators. Mathematical theories concerning physical phenomena will thus no longer be considered "absolute proof", as it happens now, especially in the so called "theoretical physics".

Mathematical proofs should be considered in physics (also in politics, economy, war, and love affairs) as hints, tips of information, advice, which have to be taken into account in physical (also in political, strategic, matrimonial, etc.) reasoning and theories. We should remember that such was exactly the attitude of great rulers of the past towards the horoscopes and conclusions derived by their mathematicians. Let us also not forget that during fourteen hundred years mathematics was providing "absolute proof" that the sun and the planets circle around earth.

Physical explanations of phenomena are rejected by mathematicians in physics ("theoretical physicists") as superfluous or even harmful wordiness. Seeking physical understanding is considered a nuisance and sign of ignorance. In natural physics, the quest for understanding is restored as a major task. So it used to be in physics, before physical explanations were replaced by calculations and knowhow.

To one who knows the meaning of *physis*, "natural physics" reads like 'soapy soap'. However, the doubling is needed, because we now have both soapless soap, and unnatural physics. Similarly, we 'forgot' that *university* means open to all, and had to organize open universities. But we shall often use the word physics singly, when it will be clear that natural physics is meant.

The name Natural Physics has some historical justification, because until the 18th century, all that was known in physics and the other natural sciences constituted NATURAL PHILOSOPHY. At those times, philosophy contained all sciences, also mathematics. Apropos, this philosophy of all science was then considered "the maid of theology". We are now facing a situation, in which some administrators consider physics a servant of mathematics.

Natural Philosophy denotes physics also in the book of Sir Isaac Newton, "*Mathematical Principles of Natural Philosophy*" (or "*Principia*", 1687), which was the "bible" of physics for 140 years. One may say that by introducing advanced mathematical methods (his calculus) into natural philosophy, Newton initiated the modern science of physics. On the other hand, it seems that by enforcing his mathematical <u>principles</u> upon natural philosophy, he opened the way of turning physics into a mathematical science.

While mathematics is good to any science, and to almost any human activity, mathematical <u>principles</u> are good to mathematics and to mathematical calculations only. Each science, as well as each human activity, has its own distinctive principles. When the principles and reasoning of another science is enforced in it, it becomes a branch of the other science, a "takeover". A natural science, governed by principles of a scholastic science, becomes scholastic itself. To survive and prosper, a science must define and keep its principles, while it may and should use the findings and methods of other sciences, but as tools, not as dominators.

6.09. Kinds of Knowledge Raised by a Natural Science

The word "science" is derived from the Latin verb *scire*, meaning "to know". We would therefore expect the natural sciences to provide a knowledge of

<u>WHAT it is</u> they DISCOVER, DISCLOSE, and deal with,
<u>WHY it is</u> the way it is, i.e., to EXPLAIN it, and
<u>HOW TO</u> adjust to it, manage, use, APPLY to make the best of it.

The three kinds of knowledge, of WHAT, WHY, and HOW TO, require different types of researchers to work at. Very seldom would a scientist be able to discover and describe a phenomenon (give the "what" of it), explain it (the "why"), and develop the possible applications of it (the "how to").

On the other hand, a scientist has to be experienced in the three kinds, because no DISCOVERY is clean of explanations and applications, APPLICATIONS cannot be developed without elements of discovery and explanation, and there are elements of discovery and application in any full EXPLANATION of natural events.

The "what-scientist" should be highly <u>observant</u>, in order to distinguish the phenomena and bodies, and <u>analytical</u>, to dismantle the observables into their components. He has to choose the governing component, "play" on it, without losing sight of the other components and factors, and be ready to "switch horses" when conditions change. He has to be able to collect enormous amounts of data, store them, and draw on them when needed. He has to describe the observables in an exact and exciting form, in order to ignite the interest of other people toward them.

We should remember that only the interest of other people can provide the funds needed for a science to prosper. Without the interest of other people, of the "organized" society or groups of it, no science and no scientist can be successful, nor will a scientist be rewarded by good publicity and pay.

The applicationist or "how to" scientist is best able to deliver what people need most, and is therefore rewarded best. He has to be a man of all seasons, able to design devices and apparatus, develop calculative methods, and be perfect in communications and public relations, especially with the potentates on whom funding and jobs depend.

To succeed, some "how to" scientists may tend to exploit the incompetence of the potentates, by selling them horoscopes, by promising to synthesize rare substances (gold, etc.), or by frightening them with dangerous forces and objects. Recent examples of such frightening inventions are "the fifth force" ("anti-gravity"), which would make the flight of ballistic missiles unpredictable, "anti-matter", which would annihilate our world, and "false vacuum bubbles", a suitcase of which would create or destroy a universe. One would be surprized to find how much tax money went and goes into the various bubbles, often due to their inflation by the media.

6.10. Infirmity of the Explanatory Work in Physics and the Bicycle Stability Syndrome

The "why" scientist, devoted to explain the phenomena of his science, is the least successful, because the interest of people in understanding is very superficial. Real in-depth explanations of natural phenomena, if any, are complicated and most often contradictory to our everyday experience and concepts. They require time-consuming intellectual efforts from the readership or audience, and cannot be put in short exciting slogans, canons, rulings, or formulae, digestible to the public and media.

What people expect when they ask their why-question is not the deep explanation, but a quicky diagnosis, slogan, or formula. The natural desire to understand is usually satisfied by answers like "because of inertia", "because of Einstein's principle", i.e., just by naming a phenomenon, a ruling, or an authority. Quite often, the desire is satisfied by a cited false, unrelated, or least related cause.

For example, the stability of a bicycle has been over a century explained "because of inertia"; more precisely, by the inertial trend of the wheels to preserve their plane(s) of rotation. But for this to be the case, the speed of bikes, especially of the lightweight ones, would have to be hundreds of kilometers per hour!

The actual stability of the bicycle is provided, mainly, by small instinctive turning of the steering handle towards the side of the expected fall, together with friction, gravitation, and the instinctive equilibrating motions of the rider's body. These are the only stabilizing factors when riding at the speed of a pedestrian, when the wheels hardly make half a rotation per second!

Such is the common picture on campuses, when the rider meets a walking classmate, who is loaded up to the throat with books and papers. The rider rides then alongside the walker for some time, exchanges information, often also kisses, without losing stability. Funny, they just left a class on the preservation of the plane of rotation, exemplified by the stability of the bicycle.

The real causes of the stability of a bicycle are not harder to explain than the irrelevant inertial trend, especially if one recalls that inertia itself is unexplained, that nobody has ever given a physical explanation of <u>why</u> bodies *have* or *exhibit* inertia. However, the preservation of the plane of rotation is "good for teaching", being both "*classroom demonstrable*" and mathematically calculable, while the instinctive actions of the rider are not.

In this "bicycle stability syndrome", which is quite common in physics, the actual causes are disregarded for the irrelevant ones, just because they seem "better" for teaching, for presentation, and for mathematical proof. Unfortunately, it is considered that all these purposes cannot be achieved without mathematical treatment, and that real physical explanations are not important.

Chapter 7

PHYSICAL EXPLANATIONS IN PHYSICS

7.01. Are There Physical Explanations At All?

In addition to cases when existing explanations are neglected, there are physical phenomena and events for which we do not have physical explanations. The question then is, WHETHER these phenomena and events are at all explainable, i.e., if their explanations are possible, but yet unknown, and will be found in the progress of science, OR that these phenomena are not explainable, and will never be, that their explanations do not exist, and never will.

These two possibilities were denoted in old philosophy by the Latin words *ignoramus*, "we don't know" (but may, in the future), and *ignorabimus*, "we shall never know".

During the last three hundred years of the development of natural sciences, more and more "unexplainable" phenomena have found scientific explanations. Therefore, it is our deep conviction and belief that all physical phenomena and events can obtain a full PHYSICAL EXPLANATION, in logical wordy sentences, and not a *quasi*, by fitting a mathematical formula or calculation; an explanation based on explainable physical facts and laws, and not by naming an unexplained phenomenon, an axiomatic postulate, principle, or an authority (a "daddy", or "big brother") who says so.

The only excusable reason for an inability of a science to explain, is its insufficient development. Galileo Galilei, who in the book *"Discorsi"* (1638) described his pioneering research of inertia and gravitation, and Sir Isaac Newton, in whose *"Principia"* (1687) we found the right quantitative laws of these phenomena, must be excused for not providing the explanation of why there is inertia and gravitation. To explain, they would need to know what became known in physics three centuries later.

7.02. Why There Are No Explanations in New Physics

When a science is sufficiently developed to explain, but does not provide explanations, then the reason can be that we either did not search deeply for them, that we were satisfied with a calculative or technological success, and accepted it as an explanation, (the "bicycle stability" syndrome), OR that we did not take into account <u>all</u> bodies, <u>all</u> factors, <u>all</u> interactions, participating in the event, i.e., that the physical model of the event is either wrong or missing.

The lack of physical explanations to most of the phenomena of new physics is also due to the lack of stimuli to search for such explanations. Phenomena of new physics are handled now by the quantum theory and by the theory of relativity in a mathematical way, based on axiomatic postulates and principles. Being mathematical, these theories can explain their mathematical operations and proceeds, but cannot provide <u>physical</u> explanations of physical phenomena.

Theories, which are unable to provide physical explanations, quite often deny their existence. This is more than just an *Ignorabimus*; it is rather the human "*Malta Yok*" symptom. The saying is ascribed to a Turkish admiral, who was leading a fleet towards Malta, but could not find the island. On return, he thus reported to the sultan that "*Malta Yok*", "*There is no Malta*".

Relativity and quantum theories provide highly complicated but successful mathematical solutions to most practical problems of their dealing. The solutions fully satisfy the needs of those who provide the funding. Hence, there is no funding, no publications, not even consideration or discussion time for explanations.

Because of the successes of mathematical solutions, the physical ways of reasoning, proof, and explanations, are now almost completely extinguished in new physics. Mathematical approaches are considered omnipotent, to such an extent that a physical relation is not considered proven if the proof is not mathematical.

Unfortunately, there is no understanding of the simple fact that mathematics cannot create physics, physical bodies or phenomena, just as it cannot create biology, life, love, money, or the economy of a country, though it may be very helpful in all these endeavors.

Mathematical theories cannot output any more physics than was put into them. If a mathematical theory or treatment is not based on the right physical model of a physical event, then the theory is intrinsically unable to provide a physical explanation of the event. Quantum theory and relativity are not based on physical models for the events of their dealing, therefore they cannot provide physical explanations for these events. To express this in a slogan, we may say that *there are no physical explanations to things which are not physical.*

7.03. Do We Need Physical Explanations and Understanding?

We perform all our bodily functions, from walking to thinking, without being able to fully explain or understand them. Most of our professional and leisure activities, like driving, using machinery, computers, TV and video, are performed without understanding the physical and other processes involved. Doctors may successfully cure their patients without real deep understanding of the illness and of the effects of their treatment.

This list seems endless. And now, if we add to it the successes of quantum and relativity theories, which proudly and religiously announce their *ignorabimus*, the inability to explain, and the non-existence of physical explanation, we should ask, what do we need explanations for, if life is so good without them.

And the answer is, that as long as things proceed well, then we do not need to explain or to understand them. Moreover, explanations and quests for understanding would retard and disturb the proceedings. So it is with walking, and with any industrial, scientific, or other activity.

However, if anything goes wrong with our walking, e.g., due to an illness or injury, then the understanding of the processes involved is crucial for the recovery. When anything goes wrong or breaks down in a production or research activity, or when a competing industrial or scientific group achieves better results, so that one is confronted with losing the market or the funding, then the understanding of the involved processes becomes urgent.

In order to have explanations ready when they are urgently needed, and thus to avoid crises, a scientific establishment must take care of the education and preparation also of the "why" scientists, devoted to the explanation of physical phenomena. There are always students ready to become "explanationists", so that the thing to do is not to suppress them, even if they become "nasty". Physics should not be treated as if it were a bible, a final, accomplished, and closed forever creation. The shy *ignoramus* has to be used much more often in teaching and presentations.

Physical explanations of natural events and phenomena are urgently needed for the dismissal of costly quasi-scientific fantasies, based on unrestrained extrapolations of mathematical results. These are the various big bang theories of how to make universes, inventions of imaginary particles and bodies, fifth forces, false vacuum bubbles, speculations about human travel with speeds close to the velocity of light, and so on. They are destructive not only because of the large amounts of funding which they absorb, but because they divert science into dead end channels, and create opportunities for charlatans to cheat people.

7.04. Educational and Social Value of Physical Explanations

Physical explanations of natural events and phenomena have great educational and social value. They are based on physical laws, models, and logic, not on mathematical and scholastic axioms, which can be turned to prove anything. They are not based on postulates, principles, and the sayings of authoritative "big brothers" in science. Therefore, physical explanations educate people to creative non-conformism, criticism, and to the request of proofs, and not to the blind acceptance of the slogans of big brothers in all branches of human activity.

This educational value, which leads to a real freedom of mind, might be an additional heavy reason why big brothers and their apparatchiks dislike physical explanations. They prefer mathematics and technology, in which one learns to obey rules, and not to explain natural events. They were ready to remove physics from school curricula, and to replace it by an "applied mathematics". But then came the post-wartime boom of the applications of physics, which made physics stay, but as an increasingly mathematical science rather than a natural one.

Under the present conditions in physics, it is very hard to educate a "why" scientist who would dedicate himself to the explanation of physical events and phenomena. In teaching, students who ask their *why's*, obtain answers which reduce to *"this is the way it is"*. Physical reasoning, logic, and proofs are replaced by mathematical derivations and unrestrained extrapolations.

Courses advertised as "physics without mathematics", or "physics for poets", as well as pop-science books, have little physics, and cannot be serious, because there is no serious physics without mathematical methods. These courses do not give physical explanations, either. Instead, they are filled with the mathematically invented bubbles, big bangs, etc., which they present with an exaltation and excitement bordering on pagan worship. The inventions are presented without their origins and methods of derivation, in a form as if they were the absolute and final truth, the divine revelation, given to us on the Mount of Sinai, and established forever.

Hence, to natural physics, the "poetical" and "pop" books are worse than texts of mathematized physics. Such texts do present the ways of derivation. This enables the creative non-conformist to at least be able to uncover the "chicken legs" on which the mathematical inventions and extrapolations stand.

7.05. Brief History of Understanding Physical Phenomena

The so called "established scientific knowledge" can explain some physical phenomena, but the explanations are usually not the full "in depth" ones. For example, Isaac Newton established the laws of inertia and gravitation, based on which he could explain the motion of planets and the stability of the Solar System. However he could not explain **why** bodies exhibit inertia, and **why** do they gravitate.

Since then, all explanations in mechanics are partial, they stop on saying "because of inertia" and "because of gravitation", without the ability to explain these two most basic phenomena. For anybody who would persist with his WHY's, Newton had the famous saying

"*Hypotheses non fingo*", "I don't make up (feign) hypotheses". This was also the closing sentence of his "Principia".

Both Galileo and Newton were flooded with questions of why there is inertia and gravitation. The word *hypotheses* in Newtons saying stands for *explanations*. The saying means that he does not make up, invent or feign explanations (hypotheses he did invent!). Newton's *non fingo* is thus less humble than a simple *Ignoramus*, which would have done better to science, but with less fame to him. Still, the *non fingo* is far more science minded than the flat "*Ignorabimus*", or "*explanations yok*" of the new physics.

The conviction in the absolute emptiness of space made Newton present light as streams of *corpuscles*, i.e., particles of unspecified matter. The corpuscles of light could propagate in the empty space without a need in a material ambient or CARRIER to carry them (now aren't these hypotheses?). Of course, a physicist would have to ask what do the corpuscles consist of, or whether they are just physically unfeasible, "as if", or "quasi-particles" of the emptiness itself. The answer was again, *hypotheses non fingo*. Though this is a no answer answer, Newton's empty space and light corpuscles survived 140 years, until 1825, when the wave-nature of light was finally established.

The wave theory of light introduced the imaginary ether as the carrier of light waves. The ether was considered a massless elastic continuous medium of unspecified matter, and was therefore physically unfeasible. Nevertheless, the continuous ether concept provided a satisfactory understanding of the then known electromagnetic and light phenomena. Based on the continuous ether concept, James Clerk Maxwell created in 1865 his successful physical and mathematical theory of the electromagnetic field and radiation.

The continuous luminiferous ether was proven non-existent by the Michelson-Morley experiments in 1887, but Maxwell's theory, and the four "Maxwell equations" are still basic for some understanding and for calculations in electromagnetism. Moreover, Albert Einstein used them in 1905 to derive his theory of relativity, which restored the empty space, and the inability to understand.

7.06. The Inability to Understand New Physics

The extent of our inability to understand Einstein's relativity can be illustrated by the following anecdote. A.S.Eddington was in the 1920s Einstein's close collaborator and interpreter. When a distinguished physicist stated that relativity is probably understood by three people only, Eddington jumped up, shouting "who is the third?".

Steven Hawking, a most prominent relativist, was more generous than Eddington. At a 1990 meeting in Jerusalem he said that modern relativity is probably understood by some tens of people, maybe a hundred. Still, his popular book on relativity (*Brief History of Time*) is selling in the ten million copies range, though nobody is able to understand its relativistic base.

The physical infeasibility of the ether concept, its breakdown, and the restoration of the empty space concept led also to the creation of the quantum theory of light (Max Planck, 1900; A. Einstein, 1905). The quantum theory (also *quantum or wave mechanics*) developed into the most powerful mathematical theory in physics, which penetrated into many branches of science.

Our inability to understand the quantum theory was best expressed by Richard P. Feynman, one of the most prominent "quantum physicists". In his 1967 paper, "*The Character of Physical Laws*", he wrote: "I think I can safely say that nobody understands quantum mechanics."

It is remarkable that though both relativity and quantum theory are based on the same model of the absolute emptiness of the vacuum space, and that the same Einstein in the same year of 1905 worked on the foundations of both, the two could not be unified into one theory, in spite of tremendous continuing efforts of the best mathematicians of the century. Each of these two brilliant mathematical theories yields successful calculative results in problems for which it was developed, and is helpless in dealing with problems of the other theory.

We may say that Feynman's statement that nobody understands quantum mechanics, should be equally applied to relativity. Eddington's and Hawking's beliefs that there might be somebody who does understand relativity is just polite or wishful thinking. It should be clear that a mathematical theory cannot produce by itself physical explanations or proofs. It can output as much physics as was put into it, and there are no physical explanations to non-physical derivations.

We will show that our inability to either unify or to understand relativity and quantum theory results from their being based on the wrong model of the empty vacuum space. With the introduction of the electron-positron lattice (epola) model of the vacuum space, all physical phenomena outside of nuclear particles can be explained, understood, and unified on the base of this model. With the acceptance of the epola model of space, all experimentally verified laws, principles, and postulates of Newtonian physics, "general" physics, relativity, and quantum theory, which were ordained without explanation, as are axioms in mathematics, turn into physically explainable and comprehensible laws.

Chapter 8
MATHEMATICAL AND SCHOLASTIC MODELS IN PHYSICS

8.01. Mathematical Modeling in Physics

When a natural event or object is presented by mathematical means only (or mostly), i.e., by numbers, mathematical signs and functions, as well as by geometric elements, then such a presentation is noted as the *mathematical model* of the event or object. A mathematical model usually disregards all or most physical processes in the events, and neglects the physical structure of the objects. It just replaces the processes and structures by mathematical functions, symbols, and geometric elements. It is based on postulates or principles which, like axioms in mathematics, do not require explanation or proof.

The mathematical model is considered valid or proven if it serves well the particular calculations and purposes for which it was introduced, i.e., if "*it works*". If it doesn't, then it is adjusted, changed and replaced until it does, without much consideration about the physics, and about the physical reasons why it does or does not work.

We often create or use mathematical models automatically, without realizing it. For example, the density of a material is defined as the quotient of the mass by the volume of any body, made of this material. By introducing the density, we automatically assume that in all points of all bodies, made of the material, the density is the same. Hence, we actually create and use a mathematical model of continuous media, in which there is a uniform density of materials, with no grains and gaps, and matter is uniform, smooth and continuous.

A mathematical model *works* only within certain physical limits, to which it is adjusted, and fails when applied to events or objects, the physical characteristics of which (sizes, velocities, densities of mass and electric charge) are beyond these limits. However, no matter how wide these limits are and how well "*it works*" within them, the mathematical model should never be considered a true and complete, "real time and real space" presentation of the event or object.

If calculative results only are required, and they are successfully provided by a mathematical model, then it is not important how remote from reality the model is, and how far it is from the natural event. But it should be clear to every user that such a mathematical model cannot provide much in the description of the event, and certainly not in its explanation.

It is a general problem with all mathematical models, that their applicability limits are not always clear or known to the users. And when the limits become known and clear, the users are so deeply involved with the model and the theories grown on it, no matter how non-physical, that it may take decades to put things right. With the Ptolemaic model of planetary motions it took centuries, and with the continuous media model we are still in trouble.

8.02. Vitality of Erroneous Scholastic Models

Scholastic models in science are not necessarily diverted from observations and experiment. To the contrary, they are usually initiated by observations, sometimes even based on experiments. However the interpretation of the observations and of the experimental results is subservient to the doctrines, dogmas, principles and beliefs of the scholar and of his scholastic system.

In simpler words, the scholars twist observed facts to make them fit the doctrines. The derived rules and procedures of a scholastic model are not as much concerned with depicting nature, as with the substantiation and glorification of the system.

Scholastic models have strong tendencies to extrapolate their findings on everything and everywhere, and to unify everything under their rules. Therefore they usually exceed their applicability limits, they lose all touch with nature, and become explicitly erroneous. However, the scholastic models insist on their doctrines and perpetuate them, by covering them up with sophisticated philosophical, mathematical, or experimental complications.

Nobody understands these complications, but would not admit it, in fear of being proclaimed stupid. So ten million people buy a book which has very little to do with reality, and in which they are, and will be unable to understand a single paragraph, which is not just a joke. And government officials spend millions to fund research programs based on such models, like the search for the fifth force.

This reminds the Hans Christian Andersen story on the new clothes of the king. Anyone who would not see them was proclaimed stupid. Therefore, everybody made believe that he is overwhelmed by the alleged beauty, texture, and colors of the clothes, and invented his own superlatives to describe them. Andersen's story has an optimistic ending, with the little child shouting that "the king is naked!". Unfortunately, such eye openers did not yet appear in physics.

Scholastic models can be philosophical, mathematical, also experimental. The greatest and most influential philosophical scholastic models were created by Aristotle (384-322 B.C.) in almost all the sciences.

8.03. Aristotle's Physics and the "Four Horse Rule"

Aristotle's work *"Physics"* served as the main source of knowledge in this science until Galileo's *"Discorsi"*, and Newton's *"Principia"*. Observations and doctrines led Aristotle to the scholastic model of motion. One of its principles was that the velocity of a body is proportional to the driving force. To illustrate this principle, he wrote that

obviously, four horses move a carriage with a velocity,
four times larger than one horse".

We shall refer to this statement as THE FOUR HORSE RULE. Both the principle and the rule sound right and are easily acceptable by students, even by working physicists.

To prove that the four horse rule is wrong, one has to try it experimentally, under different conditions. However, an observation of a fast motion, just by inertia, without any horse or "driving force", would be sufficient. It might also be sufficient to recall that there is a limit to the velocity of a horse, as well as of any animal, ourselves, our cars, boats and planes. Thus, if with one horse a very light carriage has a velocity close to the limit, of about 15 meters per second, then adding any amount of "identically" good horses cannot increase the velocity at all!

In the other extreme, when the carriage is heavily loaded, and one horse cannot make it move (velocity is zero), but two horses can (velocity is, say 1 m/s), then the doubling of the number of horses does not double the velocity, as required by the rule, but increases it an infinite number of times!

Nevertheless, there will usually be certain immediate cases when the rule works. For example, when one horse "schlepps" a heavy carriage with a velocity of 1 m/s, two horses cause a velocity of 2 m/s, and with four horses the carriage runs, by chance, with a velocity of 4 m/s, or something like that. However, such cases, when "it works", or "it fits", do not make the rule right. Here the physics alone is much too complicated for a scholastic rule to be right, even with no account of the mental and other conditions of each horse and driver.

One wonders how such simply wrong rules and models could govern science during two millennia, and be even now so easily acceptable. No doubt, Aristotle was one of the greatest geniuses of humanity, and there is plenty to learn from him even now. But he made many mistakes and misstatements, as did all geniuses before and after him. Weren't there people to see the mistakes and point them out? Or was it the fear of the subsequent condemnation, which kept them quiet?

One also wonders how much more people should learn, in order to get rid of the scholastic authoritarianism in science and life, so that relating a statement to Aristotle, to Newton, to Einstein, to Stalin, or to a guru, etc., will not make it doubtlessly right and accepted. And how much will it take until people, who are critical about such statements, and point out their erroneousness, will not be condemned, and will not have their work rejected for this "mortal sin".

8.04. Early Geocentricity and Heliocentricity

Another example of a philosophical scholastic model, related to our work, is Aristotle's GEO-CENTRIC (earth in the center) model of our planetary system. The geocentric model is based on the common observation that the celestial bodies rotate around us, and that we (particularly, the eye of the observer) constitute the center of the horizon, the center of the sphere of the sky, and thus the center of the world.

It was found in antiquity that thousands of seen stars rotate around earth without any observable change in their positions relative to one another. These *fixed stars* seem to be "nailed" on the inside of a black spherical surface, which rotates around us with a constant period of 24 hours. Contrarily, the then known five *planets* ("wandering" stars), Mercury, Venus, Mars, Jupiter, and Saturn, as well as the Sun and the Moon, are seen to not only rotate around us, but to move also relative to the fixed stars.

Aristotle's model of the planetary motions shows his deep knowledge of many particulars of these motions. The model enabled the needed calculations of the positions of planets. It corresponded to the scholastic doctrines of those times, and was accepted by all religions.

Amazingly, the same observations led Aristarchus of Samos (3rd century B.C.) to conclude that the Earth rotates around its axis and circles around the Sun. Thus, Aristarchus initiated the HELIO-CENTRIC model, with the Sun in the center of the Solar System. Though this is the physically right model, it contradicted the physically unsound Aristotle's model, as well as the ruling philosophical and religious doctrines. Therefore, the heliocentric model was condemned, together with its creator.

8.05. Practical Needs of Knowing Planetary Motions

The seen motions of planets among the various *constellations* of the "fixed" stars are highly complicated. Each planet has its own path (trajectory) and velocities, which are not constant, sometimes even opposed to the rotation of the sky and of the fixed stars. It also was (and is) very difficult to find a regularity in their seen motion.

The precise knowledge of the motions of celestial bodies was needed, first of all, for the measurements of time: the hour and minute of day or night, the day of the month, the month of the year, the count of years, and the establishment of an appropriate CALENDAR. At ancient times, of no clocks or watches, the motions of celestial bodies were the only time-keeping device for all these purposes.

The second need was for seagoers and travelers to establish their location. This need can be answered with high accuracy when celestial bodies can be observed, and if their positions at any given moment of time are known. Finally, there was the need of *horoscopes*, without which the Kings, princes, and other potentates, could not manage their political, military, matrimonial, and just everyday affairs.

To satisfy all these needs (and be appropriately rewarded), astronomers had to be able to measure, calculate, and foresee the positions of celestial bodies at any given moment back and forth in time. This led to the creation of the Ptolemaic model of the motion of planets (by Claudius Ptolemy in the 2nd century A.D.).

8.06. The Ptolemaic Model of Planetary Motions

The Ptolemaic model was a mathematical scholastic model, based on three axiomatic postulates: one, due to our observations and perception, was the geocentric postulate, stating that

the earth is in the center of the world.

Two postulates were due to the philosophical, ideological, and religious beliefs that, being a creation of the Almighty, the world must be harmonious. These two "world's harmony" (*Harmonia Mundi*) postulates state that,

all motions of celestial bodies must proceed at constant speeds;
all orbits of celestial bodies must be circular.

To account for a non-uniform or for a reverse motion, a planet was considered to move not directly on its orbit around earth but on another circle, named *epicycle*, the center of which was moving on the orbit. When the directions of motion of the planet and the epicycle are similar, then the observed combined velocity of the planet is increased.

When the directions are opposite, the observed combined velocity of the planet is reduced. At this time, if the velocity of the planet on the epicycle exceeds the velocity of the epicycle along the orbit, then a reverse motion of the planet can be observed.

Such motion of planets is physically unfeasible, but philosophically reasonable. It is mathematically perfect, because the constant speeds of the planet and of the epicycle, as well as their radii and inclinations can be adjusted to fit the observed motion. The results can then be used for all practical purposes.

Progressing trade and travel demanded more accurate assessments of time and location, so that more accurate instruments were built for astronomical measurements. When discrepancies were detected between the calculated and observed positions of a planet, then an additional circle, a hypocycle was introduced. The planet was then considered to be circling on the hypocycle, the center of the hypocycle was circling on the epicycle, and the center of the epicycle was circling on the orbit around earth.

If this did not fit the observed motion, an additional circle was introduced, the hypo-epicycle. The planet was then circling on the hypo-epicycle, the center of which was circling on the hypocycle, the center of which was circling on the epicycle, the center of which was circling on the orbit around us.

In this manner, any degree of accuracy of calculations can be achieved by adding a cycle. However, the calculations in this "ADD A CYCLE" game become very cumbersome.

8.07. The Copernican Heliocentric Model

By 1507, Nicholas Copernicus (1473-1543) found that the reverse motion of planets would disappear from calculations if one assumed that earth is circling around the sun. Copernicus was arguing that by the mathematical consideration of the motion of one body, the Earth, one gets rid of having to consider many motions: the reverse motions of all planets, and of their epi-, hypo-, and hypo-epi-cycles. This mathematical simplification convinced Copernicus to restore the ideas of Aristarchus and create the mathematics of the heliocentric model.

In his mathematical presentation, Copernicus preserved the erroneous "world's harmony" postulates of circular orbits and constant speeds. Hence, though based on the physically right heliocentric model, his presentation could not yield accurate calculative results. Thus, the Copernican model did not provide the needed advantage to successful astronomers or position seekers.

Therefore, the manuscript *Commentariolus*, which he started to distribute among mathematicians by 1514, and the book "*De Revolutionibus...*", or "On Rotations of Celestial Bodies" (1542), obtained a cool reception by the scientific community.

Being mathematical, the Copernicus' works were not opposed by the church (until several decades after hid death). But the inability of the Copernican model to provide accurate calculative results made it impractical. This was the main reason why it was not accepted, and opposed by astronomers during two centuries.

Nevertheless, some astronomers, as well as a few philosophers were impressed by the Copernican Model. This forced the astronomers of the establishments to improve their measuring apparatus and raise the accuracy of their observations in order to disprove the model, and heliocentricity with it.

One of them was Ticho Brahe (1546-1601), the astronomer of the Danish royal court. He convinced his master to provide funds, and built the most precise *astrolabia* of all times. With it, he succeeded to collect an impressive amount of data on the motion of planets. However he had no time left to process the data.

8.08. The Keplerian Model of Planetary Motion

Ticho Brahe's student and co-worker Johannes Kepler (1571-1630) inherited all results of his teacher's measurements to carry-on the mission of disproving heliocentricity. However, after years of trial (Kepler, too, was making a living by selling horoscopes!) he found that all data fit best and prove the motion of planets around the sun, though
  not on circles, but on elliptic orbits, and
  with speeds which are not constant but vary,
depending on the position of the planet on the orbit.

In the book "*De Harmonia Mundi*" (1609), Kepler dismissed the geocentric postulate as well as the "world's harmony" postulates, and derived the physically feasible and correct heliocentric laws of planetary motion.

Kepler's model and Laws are mathematical; they do not consider the physical interactions, causing the planets to move as they do. These interactions are due to the phenomena of gravitation and of inertia, the laws of which were derived by Newton seventy years later, and published in his "*Principia*" (1687).

Kepler's Laws provide physical feasibility and mathematical accuracy. However, they require more progressive mathematical approaches (elliptic geometry, accelerated motions) than the Ptolemaic calculations. Even the most cumbersome Ptolemaic calculations, with many circles, but of constant radii, and with many speeds, but of unchanging values, were more familiar and domesticated. Moreover, it was simpler to adjust an observed discrepancy with the "add a cycle" game, than to bother with elements of elliptic motions.

As a result, even two centuries after Kepler, astrologers seeking high accuracy, continued to use the Ptolemaic model. Nowadays, the difference between heliocentric and geocentric calculations is reduced to split seconds of computer time. Therefore, if calculations are required, then it is not significant whether you accept the sun as the center of the Solar System, or your own "private" eye as the center of the world.

Nevertheless, one should strongly keep in mind, that the outburst of Newton's physics, the discovery of the planets Neptune and Pluto, as well as our space activities, would not be possible if, instead of Kepler's Laws, the <u>mathematically exact</u> Ptolemaic model would continue to be the "established scientific knowledge" as it was during fourteen centuries.

Even with the use of the most powerful computers, one would be unable to compile a physical background of interactions which would cause planets to move according to the postulates and procedures of the Ptolemaic model, nor is it possible to find physical explanations for it. There are no physical explanations to things which are not physical. The model is simply erroneous, and celestial bodies do not move according to it, no matter how well the model describes their motions and how exactly it allows the motions to be foreseen.

But even the calculative success of the Ptolemaic model is limited to the few planets for which it was developed. With the progress of astronomy, somewhere the "add a cycle" game became ineffective, and the model became a dead end street, in spite of the credibility of our perception of being the center of the world.

8.09. Conclusions Concerning Mathematical Models

The review of the Ptolemaic model, which dominated science during fourteen centuries, and the reviews of the mathematical models, which dominate in physics now, like the quantum and relativistic models and theories, lead to conclusions, as follows.

1. Erroneous scientific concepts are not necessarily based on surrealistic inventions (but may lead to them!). On the contrary, erroneous concepts are usually based on healthy perceptions, observations, experiments, and calculations. However, some of these bases, or all, are misinterpreted (it's easy!) to fit a prejudiced ideology, or scholastic dogmas.

2. Prejudices and scholastic dogmas have tremendous vitality. Therefore, the erroneous scientific concepts, which are based on them, seem highly convincing to everybody, just "natural", and survive for decades after having been doubtlessly disproved.

3. If a scientific model yields correct calculations for observed phenomena, but is unable to explain their physical or other natural causes after decades of use, then some of its fundamental concepts and postulates must be erroneous.

4. Models which are unable to provide physical explanations tend to cover up for it by claiming that explanations do not exist. This reminds the *"Malta Yok"*, "There Is No Malta" report of the Turkish admiral who could not find the island. The cover up is also a conceited replacement of the shy *"Ignoramus"*, "we don't know", by the impudent *"Ignorabimus"*, "we shall never know".

5. The calculative successes of a mathematical model are limited to those phenomena and objects for which it was developed. With the discovery of new phenomena, the model becomes less and less helpful, until it shows up as a dead end street.

6. When a mathematical model turns into a dead end, mathematicians try to revive it by using the "ADD A DIMENSION" game (up to 502 fictitious dimensions added now to the 4 real ones), the

"ADD A PARTICLE" game (already a hundred of fictitious particles "created"), and the "add a force" game, all analogous to the "add a cycle" game of the Ptolemaic model. The so created new offshoots initially seem like through roads, but sooner or later they, too, show their dead ends.

7. Mathematical models and theories which are unable to explain the physics behind their postulates cannot be unified. Each has successes with its group of phenomena and objects (though loudly and repeatedly claiming universal applicability), and is unsuccessful with phenomena dealt with by the other theory.

8. Disclosure of the erroneous foundations and postulates reveals also the reasons for the inability to provide physical explanations. It usually is that not all relevant bodies, entities, and processes (not all "members of the cast") were accounted for. And those which were, just do not behave and proceed in the way assumed by the model or theory.

CIRCLE THREE

Chapter 9.

ASSESSMENTS AND INTERPRETATIONS OF EXPERIMENTS

9.01. The Direct Results of the Michelson-Morley Experiment

The straight and direct result of the Michelson-Morley experiments is that the velocity of light, propagating in the direction of Earth's motion around the Sun, is the same as the velocity of light in directions opposite and perpendicular to the motion of Earth. This direct result means that *light, emitted by Earth and earthly bodies*
       (*i.e., by bodies of atomic matter*),
propagates in our space with a velocity,
which is not influenced by the motion of the emitting body.

Any formulation of the results has to keep and accentuate the limiting conditions of the experiment, which are,

    1. The light is emitted by an atomic body. We do not know if the results hold also for light emission by bodies of nuclear matter, like a Supernova star, or by bodies of yet undetermined kinds of matter.

    2. The result holds for light propagating in OUR SPACE, which is the surrounding space of Earth. It is possible, that the same result would be obtained throughout the Solar System. However, we do not know what is happening in regions very close to the Sun, or around other, much more massive stars or bodies of atomic matter, let alone in space around bodies of nuclear or other matter.

    3. The velocity of the emitting atomic body RELATIVE TO THE SURROUNDING SPACE is around 30 kilometers per second, which is the average velocity of Earth's motion around the Sun. In the Solar System, the largest observed velocity of an ATOMIC body relative to its surrounding space, e.g., of a comet, is ~60 km/s. It is possible that the same result would be obtained also at such

velocities, or up to, maybe, a 100 km/s. But it would be very inappropriate for a scientist to generalize a result any further (except if he believes in Aristotle's "four horse rule").

Keeping in mind these limitations, we may generalize the results of the Michelson-Morley experiment, to say that,
>   the velocity of light, emitted in our space
>   by a moving atomic body,
>   is independent of the velocity of the body.

The direct result leads to the <u>secondary result</u>, that
>   the moving light-emitting atomic body does not push or pull
>   the carrier in which the light is propagating,
>   nor does the motion cause winds or currents in this carrier.

The tertiary result is then, that the carrier of light cannot be any CONTINUOUS substance, because motion in a continuous substance might have caused winds or currents in it.

9.02. Misinterpretations of the Michelson-Morley Results

A.A.Michelson and E.W.Morley could not know in 1887 (nor could A.Einstein in 1905), that all their apparatus, producing and measuring light, and the Earth as well, are actually networks of nuclei and electrons at large distances from one another. They considered all bodies, as well as the "luminiferous" (light-carrying) ether as continuous media. Therefore, it was absolutely clear to them that the motion of the allegedly continuous and impenetrable Earth through the continuous ether must cause winds or currents in it.

From the absence of any pushing, pulling, or winds in the carrier of light by the motion of Earth, they derived that there is no ether. The right conclusion would be, that whatever it is that carries the light, cannot be continuous. Unfortunately, all matter, people could and can think of, was considered continuous. Therefore, the absence of a continuous ether was (and still is) understood as the absence of any material carrier of the light waves. This caused plenty of confusion and distress in physics and in related sciences, and still does.

To cope with the so misinterpreted results of the Michelson-Morley experiments, A.Einstein in 1905 introduced, among others, his "second postulate of special relativity", stating that, *"light is always propagated in empty space with a definite velocity c which is independent of the state of motion of the emitting body."*

Comparing with our formulation (in Section 1) of the direct results of the Michelson-Morley experiment, one might say that the wording of this postulate is quite cautious. However a deeper insight reveals some words, like "always", "light is propagated" (instead of "propagates"), "definite velocity", implying that the velocity is finite, for "always" and for everywhere. This wording leads away from the experimental results, and prepares a foundation for the non-physical character of Einstein's relativity, with its statements that,

    the vacuum space is absolutely empty of any matter;
    there is not ANY material carrier of light whatsoever;
    the velocity of light c, is always and everywhere the same,
and is constant, in this universe, and in all others to come.

None of these postulatory statements was ever proven experimentally. Their "proof" rests on some satisfactory results provided by relativistic calculations. Thanks to them, and to the exterminating actions of the physics regime against any alternatives, there were already three generations of physicists brought up and educated in the deepest belief that relativity with its subsequent big bangs is the whole and only absolute truth.

Another non-physical and anti-scientific act is the postulation of the <u>universal</u> constancy of the velocity of light. It does not follow from the Michelson-Morley experiments, nor was it ever proven by any experiment. What is true is that light, from wherever it might come to us, always reaches our detecting apparatus with the same velocity c, equal to 300,000 km/s. But this is so because around all our detecting apparatus, space is one and the same. Therefore, the velocity of light, which depends on the physical conditions in space, is also the same.

Entering our space, light propagates with the velocity c, corresponding to the conditions here. Light is not "propagated" by the conditions in the other regions of space, in which it was emitted, and which it had to pass on its way to us. Entering our space, or any other region of space, light "does not remember", and is not influenced by the velocities which it had before, and propagates with the autonomous velocity, corresponding to the physical conditions in this region.

9.03. The Rutherford Experiment

When Ernest Rutherford started his experiments with the scattering of alpha-particles in thin foils of metals (silver, etc.), atoms were considered as droplets of positively charged liquids, with the negatively charged electrons floating in them. This *Liquid Droplet* model of the structure of atoms agreed with the beloved philosophical concepts of the continuity and the unity of matter, because the liquid of which the atom consisted was certainly continuous, and its density was obviously comparable to the density of the material.

Because the liquid was believed to "fill up" the atom and be distributed uniformly over its entire volume, Rutherford expected that the impinging alpha-particles will meet atoms and be scattered by them at moderate or small angles. Though the foils were very thin, they were still thousands of atoms thick. Hence, the probability that an alpha-particle will make its way through the foil without being scattered was calculated to be extremely small.

The results of Rutherford's experiments, published in 1911, were astonishing. It turned out that almost all alpha-particles penetrated through the foils without significant diversion. Only a very small number of particles was scattered (one in many thousands), but under very large angles, up to 180 degrees (i.e., backwards).

The physical interpretation of these results is that almost the whole volume of the atoms is penetrable to the alpha-particles, or transparent to them. The alpha-particle has mass and volume almost 8000 times larger than an electron. Therefore, the alpha-particles, impinging upon an atom, do not "see" the orbital electrons, and move through the atom as if it were completely empty.

However there must be in the atom a spot, or NUCLEUS, of very small volume. When this spot is hit by an impinging alpha-particle, then, and only then can the particle be scattered as strongly, even backwardly, as observed. For the nucleus of the atom to be such a strong scatterer, the densities of mass and of positive electric charge in it must be comparable with the densities of mass and charge in the alpha-particles.

Rutherford's experiments allowed precise measurements of the numbers of scattering events, of the angles of scattering, and of the velocities of the impinging and scattered particles. This data enabled the calculation of the masses, sizes, and densities of the nuclei and alpha-particles, and the establishment of the PLANETARY structure of atoms.

The experiments made it clear that all the mass and positive electric charge of an atom is concentrated in its nucleus, the diameter of which is hundred thousand times smaller than of the atom, and the volume is a quadrillion times smaller.

The negative electric charge of all orbiting electrons in the neutral atom is equal and opposite to the positive charge of the nucleus. However the combined masses and volumes of the electrons in the atom are negligible compared with the mass and volume of the nucleus. Hence, only a quadrillionth of the volume of the atom, and of atomic bodies in whole, is occupied by particles. The rest of the volume, i.e., almost all the volume of atoms, is as empty as space.

9.04. Unpublicized Consequences of the Rutherford Experiment

Rutherford's most important achievement, derived from his experiments, is the establishment of the planetary structure of atoms. However, Rutherford could not explain why only certain electron orbits in the atoms are stable. The condition for this stability was derived by Niels Bohr two years later (in 1913), but remained unexplained in standard physics till this very day.

In spite of this deficiency of Rutherford's planetary model of the atom, due credit is given to him, and the model was and is duly publicized. However, several other important direct consequences of Rutherford's Experiments are not publicized, even ignored or silenced. These consequences are, as follows.

1. The intrinsic non-continuousness of all atomic matter;
atomic matter is built of discrete "grains", atoms and molecules, between which there are some distances, and atoms consist of nuclei and electrons, the distances between which are hundred thousand times larger than the diameters of these particles.

2. The space-like "emptiness" of all atomic bodies;
this includes us, and Earth, and all our possessions; only a quadril-

lionth of the volume of atomic bodies is occupied by the nuclear particles of which they consist. The rest of the volume, i.e., almost all of it, is just space, or as empty as space is postulated to be.

3. The transparency of atomic bodies to nuclear particles; this includes alpha particles and alike, but especially the neutral particles, neutrons, electron-positron pairs, or neutrinos. An atomic body can thus move through collectives of such nuclear particles without significant resistance, without making winds in the collectives, but waves.

It is very hard to understand, how it is possible that these important consequences of the Rutherford 1911 experiments were not publicized and did not find their way into science and education. Obviously, the domineering theories in new physics would be dangerously shaken, if atomic bodies and the radiation-carrying space were considered networks or lattices of nuclear particles. Hence the stubborn sticking to the continuous media model, far beyond its applicability limits. This caused many troubles to physics until the present day, and as things look now, will continue to do so for many years to come.

The unpublicized consequences of the Rutherford Experiments make it clear that the existence in space of a rare lattice of nuclear particles is in full agreement with the results of the Michelson-Morley results, while the Anderson Experiment proves, that this lattice should consist of electrons and positrons.

9.05. Physical Interpretation of the Anderson Experiments

Carl David Anderson discovered in 1932 a particle, having the positive charge of the proton, and the mass of the electron. The particle, named positron, is therefore a "sister-particle" of the electron; both particles have equal masses and equally opposed electric charges which can neutralize each other.

Anderson's experiments showed also, that when a gamma-ray energy amount, or quantum, of no less than 1.02 million electron·volts (1.02 MeV, or 160 quadrillionths of a joule, or 39 quintillionths of a calory) is absorbed in any point of space, a free electron and a free positron emerge out of this point.

Inversely, when a free electron meets a free positron, the two nuclear particles may disappear in a point of space, out of which will then emerge two (at least) gamma-ray quanta, of combined energy equal to 1.02 MeV or 39 quintillionths of a calory.

The Anderson experiments, when interpreted without mathematical and philosophical prejudices, prove that the vacuum space contains electrons and positrons, bound in space by the energy of 1.02 MeV per this pair of particles. As long as the particles remain bound in space, they are not easily detectable.

When the bound electron-positron pair absorbs this amount or *quantum* of energy, the two particles are freed out of their bonds in space, and can be detected by our apparatus. Hence, 1.02 MeV is the binding energy of an electron and positron in our space. Inversely, when a free electron and a free positron happen to meet, they may be caught into bonds in space, and disappear to our detecting apparatus. Then their binding energy is released, or *emitted*, usually in the form of two (at least) gamma-ray quanta.

9.06. Anderson's Experiment and the $E=mc^2$ Formula

It is very unfortunate for the natural science of physics that in the Anderson Experiment times, the mathematical theory of relativity was already accepted, although not yet dominant in physics. Nevertheless, the appearance and disappearance of electron-positron pairs was interpreted as their creation out of emptiness and their annihilation into the emptiness, in accordance with Einstein's $E=mc^2$ formula for the alleged equivalence of mass and energy.

Hence, Anderson's meaningful physical experiment, which is one of the pillars, establishing the physical structure of space, was misinterpreted to fit and glorify a mathematical model, opposed to the physics of space.

However the $E=mc^2$ formula has nothing to do with making mass from energy or energy from mass. This is physically just as impossible as *making (not buying!)* a loaf of bread from a dollar bill, or making a dollar bill from a loaf of bread (*making, not selling or exchanging!*).

The physical meaning of the E=mc² formula is that if an energy E was absorbed in space (or in a nuclear reaction), resulting in the appearance of a mass m of electrons, positrons, or other nuclear particles, then the absorbed or disappeared energy E can be calculated by multiplying the appeared or freed mass m by the square of the velocity of light c. Particularly, dividing the absorbed energy E by the number of freed particles, or freed electron-positron pairs, one obtains the binding energy of the particle, or of the electron-positron pair.

9.07. Analogy: Creation and Annihilation of Free Ions in Crystals

Processes similar to Anderson's "creation and annihilation" of electron positron pairs occur also in solids. For example, a crystal of table salt (sodium chloride) consists of positive sodium ions and of negative chlorine ions. Their charges are equal to the charges of the positron and the electron, and they neutralize each other in the crystal. As long as the ions are bound to each other in the crystal lattice, we cannot detect them with simple mechanical or electrical means.

But when ultraviolet radiation of quantum energy 8 electron volt is absorbed in the crystal, then there is a pair of sodium and chlorine ions freed in the crystal per each such absorbed quantum. The freed ions can then be detected by the electrical conductivity which they cause in this insulating crystal, by the coloring which they cause in this perfectly transparent crystal, and by other effects.

The freeing of ions out of their bonds in the crystal lattice can be quantitatively described by the formula $E=mv^2$. Here E is the absorbed (or emitted) radiation energy, m is the mass of the freed (or re-bound) ions, and v is the velocity of sound in the crystal. The formula describes also the reversed process of capturing free ions into the lattice bonds, with the emission of their binding energy, in light quanta of total energy 8 eV per captured pair.

We know well that 8 eV of energy cannot create or make the pair of sodium and chlorine ions, but can only free the two ions from their bonds in the crystal lattice. Exactly so, 1.02 MeV cannot create or make the electron-positron pair, but can free the two particles out of their bonds in the lattice, which they form in our space.

9.08. The Inability to Make or Destroy an Electron or Positron

In the 1930s, the only available source of high energy quanta were the cosmic gamma rays. With the construction of high power accelerators, particles of multi-MeV energies were obtained, but they were unable to create or annihilate a single electron or a single positron. And now we are able to bombard the "empty space" with particles of a million MeV energy. But even they are unable to create or destroy a single electron or positron, while one MeV is sufficient for the <u>appearance</u> of an electron and a positron out of the emptiness.

It should therefore be absolutely clear to anybody with a physical approach to natural phenomena, that
>electrons alone and positrons alone
>are not creatable and not destructible,
>at least not out of, or into an empty space,
>and not with all the now available million MeV energies.

Therefore a 1.02 MeV cannot and does not create an electron and a positron, if even million times higher energies cannot create one of these particles. The appearance of an electron and of a positron out of space, due to the absorption of the 1.02 MeV energy, is therefore proving beyond any doubt that these particles exist in space, bound to it by this amount of energy.

The unresisted motion of atomic bodies in space is then possible, if the distance between the bound particles in space is much larger than the diameters of these particles. The distance should also be sufficiently large and flexible to enable the nuclei of the atoms of the moving body to squeeze their way between the bound particles.

Chapter 10

THE ELECTRON POSITRON LATTICE (EPOLA) STRUCTURE OF SPACE

10.01. Space and the Fields of Forces

In every spot of space around (and in) us, there is the action of the GRAVITATIONAL FORCE on an inserted body. The action can be perceived with our senses, as heaviness, or weight. It can be measured with a spring balance, or other instruments. In physics, *the space, or region of space, in which there is the action of a physical force, is the* FIELD *of this force*. Hence, the surrounding space is the field of the gravitational force, or the GRAVITATIONAL FIELD.

In every spot of space around an electrically charged body, or around a magnetized body, there is the action of the electric force, or of the magnetic force. These actions can be revealed (detected) and measured, when an electrically charged body, or a magnetic body, is inserted into any spot of such space. Hence, this space is an ELECTRIC FIELD, or a MAGNETIC FIELD, correspondingly.

Because the fields of the physical forces can be revealed, detected, and measured, we should agree that *fields of physical forces exist in nature*, and not only in our imagination or presentation. Hence, the fields of physical forces, i.e., the electric field, the magnetic field, and the gravitational field, are physical entities.

The fields of physical forces exist and act in space, not only when the space contains bodies of atomic matter, but also when it is evacuated (pumped out) of them. This VACUUM SPACE is therefore the CARRIER of these forces and fields. Moreover, this space carries tremendous amounts of ELECTROMAGNETIC RADIATION, traveling (propagating) from their sources, during split seconds, or during tens and millions of years. Therefore, space is a physical entity, and always has to be treated as such, though the physical structure of space may not yet be accepted by the *scientific community*.

10.02. The Physical Structure of Space

Results of all related experiments, when interpreted without the influence of past authorities and of philosophical prejudices, prove that the vacuum space is not completely empty. It contains particles of nuclear matter, - electrons and positrons, which are forming a widely spaced network or LATTICE of material points.

The lattice of electrons and positrons in space does not resist the motion of atoms and atomic bodies, because of three reasons. First is the extremely small radius of these particles (no more than a tenth of a fermi (0.1 femtometer) or a tenth of a quadrillionth of a meter. Second is that the distance between two neighboring particles in the electron-positron lattice is 50 times larger than their sizes. Third is the very large distance between the electrons and nuclei in atoms.

With our bare senses we are unable to detect single atoms and molecules. We do not even detect the 'regular' air! We sense the air only when it is windy, or polluted, too hot or too cold. No wonder that we cannot sense particles of nuclear matter, which are millions of times smaller than atoms and molecules.

The electrons and positrons in the space lattice are strongly bound to it. Therefore, they cannot be "pumped out" by the best vacuum pumps we can think of. They can be detected as individual particles or particle pairs, only after being torn out or freed off the lattice, by submitting to them their large BINDING ENERGY.

When disturbed by energies smaller than the binding energy, the electrons and positrons of the space lattice cannot be detected individually. But the disturbance displaces the particles from their positions in the lattice. The displacements cause DEFORMATIONS, VIBRATIONS, and WAVE MOTIONS in the lattice.

The various deformations of the lattice are detected as gravitational, inertial, electric, and magnetic forces. The various wave motions, due to the vibrations of the electrons and positrons in the lattice, are detected as various kinds of ELECTROMAGNETIC RADIATION. One of these kinds is the "visible" radiation, or light.

10.03. Presentation of the Epola Model of Space

The ELECTRON POSITRON LATTICE, or EPOLA, for short, model (or theory) of the physical structure of space, was first presented in 1973, and submitted since then in a hundred contributions to journals and scientific meetings, as well as in a review book. The book "*The Electron Positron Lattice Space*", had limited first printings in 1987, 1988, and 1990. It was distributed among working scientist, and only a few copies found their way to libraries.

The epola model was never proven wrong, but was not accepted by the *scientific community*, yet, only because it contradicts some parts of the theories which are now ruling in physics.

Wrong theories survived in science during centuries and millennia, and the acceptance of the right but contradicting models took decades and centuries. Therefore, the non-acceptance of the epola model should not be considered as a proof, not even as an indication of whether the model is right or wrong. This book is meant to convince the reader that the epola model is right, and also much more physical, human and humane than the ruling theories.

The theories which are now ruling in physics, particularly relativity and quantum theory, are based, like mathematics, on axiomatic unexplained postulates and principles. These theories treat natural phenomena as if they were mathematical items, and are concerned with finding mathematical solutions to physical problems, even by twisting and dismissing physical laws.

To derive the solutions, the ruling theories construct mathematical models, inventing tens and hundreds of imaginary dimensions (502 of them "sold" by 1985), hundreds of imaginary particles, and mathematical functions (operators), which "create" or "annihilate" dimensions, particles, and whole universes, just by being pencilled or keyed down. Thus, they turn physics into a scholastic mathematical science, bordering with science fiction, and containing more fictions than science.

The solutions to physical problems, provided by the ruling theories, are highly complicated, operable only by insiders. However, these solutions satisfy the immediate practical needs of people, who work in physics and depend on it. When they do not, the theorists invent some other mathematical models and operators, or change the aims of the existing ones, until a satisfying solution is obtained.

Such is the beneficent power of mathematics. Not being a natural science, mathematics is not bound to depicting natural phenomena as they are. Therefore, mathematical presentations can be tailored and directed as you wish, if you only know how.

The ruling theories do not care whether the mathematical models are physically feasible, neither do they care about providing physical explanations of the phenomena they deal with. Actually, being mathematical, these theories are unable to provide such explanations. Therefore, they either deny the very existence of physical explanations and proofs, or present as such some mathematical derivations. At best, they name an axiomatic, unexplained postulate or principle as the cause of events.

Based on the epola model of space, all working formulae of relativity and quantum theory are derived using simple algebra. The working postulates and principles of these theories obtain their physical explanations. They thus cease to be divine revelations, and turn into regular understandable laws of nature.

Based on the epola model, different mathematical games (add a dimension, add a particle, add an operator), and the twisting of the concepts of time and size, the "twin effect", etc., become superfluous. Mathematical exaggerations, illegal extrapolation, and other tricks, which lead to so many fictitious results, are proven wrong.

As examples of exciting but fictitious results of mathematical exaggerations we may mention our allegedly ever exploding universe, created in a split second *big bang*, suitcases of *false vacuum bubbles*, which can make or destroy a universe, and the travel with velocity close to the velocity of light, which can keep the traveler young forever, and even younger than at the start of the voyage.

Imagine replacing the slogan "Join the Navy and See the World", by "Join the NASA and See the Universe". Then come back after a few years, just in time to marry the youngest cousin of your twin brothers grandson, as promised by the relativistic twin effect. With such a scenario for a best-seller, why would one need real physics?

With all the advantages listed, the epola model is also the only one which provides full <u>physical</u> explanations of all observed **out of nuclei** physical events and phenomena. Thus, the epola model is more reliable, and far better and easier to understand, to teach, and to apply in technology and calculation, than the presently "established" and ruling theories.

10.04. Our Physical Concepts of Time and Dimensions

Time is considered in normal life to run or flow smoothly and uniformly, independent of what we are doing in it, whether we work or dance, move or rest, and whether our clocks and watches work. Even if Earth and planets stop rotating, our time would be running exactly as now and before, at the same rate of flow, except that it would be harder to measure, by whoever would be there for it.

Such is our understanding and concept of time, shared by a great majority of people. One may consider time a coordinate or dimension, but then our time is the only absolutely independent coordinate. Because we do not worry about saving the ether, or the absolute emptiness of space, or any other fiction, we do not have to speculate with this most basic coordinate, and shall keep it as is, in all our proceeds.

From our millennia long experience we know, that the dimensions of a body change only when there is an internal or external action on it. To change the length of a body, one should heat or cool it, compress or pull it, cut or ground or etch it, and so on. If the length of a body would shorten in motion, then this could occur only because of a physical or chemical interaction between the body and the medium in which it moves. This is our concept of the dimensions of a body, shared by a great majority of people. Because we do not have to save the ether, or the absolute emptiness of space, or any other fiction, we shall keep this concept as it is, in all our proceedings.

In our natural system of orientation, time is running or flowing at a constant rate, absolutely independent of what is happening in time, or what we are doing in it. Time runs at the same rate, independent of whether the Earth is rotating around its polar axis and around the

Sun, or whether the Earth rests, and everything is rotating around Her. Our time flows at the same constant rate whether we travel with the highest speed of 60 km/s (220,000 km per hour), achievable in the Solar System, or watch the voyage on TV at home.

Hence, if considered a co-ordinate, in which to order events as they occur, our time is the only absolutely independent coordinate. Everything is changing, developing and disintegrating, depending on time, but time changes at its constant rate, independent of anything.

Our natural system of orientation in space is based on three lines only, or three coordinate axes, along which we measure the three dimensions of any region of space or of physical bodies. One of these lines we may direct to the right and left from us, or from any other point of ORIGIN. This line is the coordinate x-axis or abscissa, along which we can arrange to measure the dimension of LENGTH.

The second line is directed perpendicular to the abscissa, say, forward and backward from us, or from the origin. This is then the y-axis, along which we can arrange to measure the dimension of width. The third line is directed perpendicular to both the x and y axes, and is pointing up and down. This is the z-axis, along which we can arrange to measure the dimension of height or depth.

Because we do not have to adjust our proceeds in natural physics to the fictitious ether, or to the absolutely empty space, or to any other fiction, we keep our basic orientation system in space and in time as it was established, and as is, i.e., with the absolutely independent time coordinate, and the three space coordinates, depending on time.

10.05. The Physical Meaning of Einstein's $E=mc^2$ Formula

Einstein derived this formula from a calculation of the electromagnetic radiation energy of an electron moving in the electromagnetic field. Unfortunately, he generalized his results, derived for electrons only, on any material point, and on *"ponderable masses as well"*, i.e., on bodies of atomic matter, by saying:

"We remark that these results as to the mass are also valid for ponderable material points, because a ponderable material point can be made into an electron (in our sense of the word) by the addition of an

electric charge, no matter how small."

Well, it is not necessarily so, not at all so. Firstly, electrons are a quadrillion times denser than atomic bodies, so that in no sense can an atomic body be made into an electron. Secondly, the density of electric charge in the electron is a quadrillion times higher than in an ion, in which it is a million times higher than in the most strongly charged atomic bodies. Hence, there is no way by which we could dream to add to a "ponderable material point" an electric charge reminding the dense charge of an electron.

The striking structural differences between the dense nuclear particles and the almost empty atomic bodies were not known by 1905. However, even with the *Liquid Droplet* model of the atom, such an automatic extension on atomic bodies of results, obtained for electrons, was improper. Provided that physical thinking was not overwhelmed by mathematical formulations and by the eagerness to extrapolate them to fit everything, everywhere.

Einstein also unjustly equated the electromagnetic radiation energy, for which he made the derivation, with any other energy, by saying:

"The fact that the energy withdrawn from the body becomes energy of radiation evidently makes no difference, so that
The mass of a body is a measure of its energy content; if the energy changes by E, the mass changes in the same sense by E/c^2."

Well again, it is not necessarily so, not at all so. Firstly, it is well known in physics that different KINDS of energy are not equivalent to one another. Some kinds are "more equal" than the others. For example, mechanical energy is "more equal" than thermal energy. Thus, only a small part of the thermal energy of a body can be converted into mechanical energy. The energy of electric currents is most versatile, but only a small part of it can be converted into radiation energy, which is the "most equal" kind.

By saying that the kind of energy "*...evidently makes no difference*", Einstein evidently played on the common misunderstanding of the law of energy conservation. This most important physical law states that **if and when** energy is converted from one kind into another kind, then the converted <u>amounts</u> of energy are equal. However the different <u>kinds</u> of energy are not equal and cannot be replaced by one another, not in life and not in nature.

It is important for a physicist to remember that mathematical formulae contain **amounts** only. Therefore, if the left-hand side of an equation is the amount of one kind of energy, and the right-hand side represents the amount of another kind of energy, then the equation equates the amounts, and not the kinds. No mathematical equation can equate <u>kinds</u> of energies, matter, forces, or other physical magnitudes, simply because the KINDS of magnitudes do not appear in the equation. What appears, and is equated, are the AMOUNTS of these magnitudes.

Contrary to Einstein's statement, in his equation, m is the change in the amount of mass of electrons and other nuclear particles, present, appearing, and disappearing in the electromagnetic field. This amount of mass is equal to the amount of absorbed or emitted electromagnetic energy E, divided by c^2, which is the squared velocity of light.

However, the equality of these amounts, $m=E/c^2$, does not mean that mass and energy are equivalent and replaceable. By analogy, if you can buy a loaf of bread for a dollar bill, then this does not mean that these two items are equivalent and replaceable. Equality and replaceability would mean that one can make or create a loaf of bread out of a dollar bill, or make a dollar bill (a real one!) out of a loaf of bread. Make, not buy, sell, or exchange!

Physical magnitudes *per se* do not appear in equations, only their amounts. Hence, the equation cannot and does not equate or replace mass by energy, or energy by mass. Therefore, the presentation of Einstein's equation as a universal "mass-energy equivalence", applicable to any body and any energy, is improper. To say the least.

The Anderson experiment is connected with the $E=mc^2$ formula insofar as if an energy E is absorbed in space, then this energy, divided by 1.02 MeV, or by 39 quintillionths of a calory, yields the number of electron-positron pairs, freed out of their bonds in space; the mass m of all these freed nuclear particles is then equal to E/c^2.

Vice versa, if a mass m of nuclear particles was caught into bonds in the vacuum space, (or into a nuclear reaction) then an energy E, equal to mc^2, is emitted in the form of quanta of electromagnetic radiation energy. The physical reason for it is easily derived when the physical structure of space is established.

10.06. The EPOLA Structure of Space

It was proven by the Rutherford experiment that atomic bodies consist of nuclei and electrons, the distance between which is a hundred thousand times larger than the sizes of these particles. Hence, submicroscopically, from the "point of view" of nuclear particles, atomic bodies are not solid impenetrable continuous walls, but networks or lattices of nuclear particles (nuclei and electrons), located far apart from one another. For example, if we would represent the size of a nucleus by one millimeter, then the distance to the next nucleus would be close to hundred meters!

Space, even when completely evacuated of atoms and molecules, may still contain the million times smaller particles of nuclear matter, which our best vacuum pumps cannot pump out, and our simple apparatus cannot detect. If the distance between these nuclear particles is much larger than their sizes, then they, too, may form in space a network or lattice of particles, far apart from one another.

Thus, the unresisted motion of an atomic body through space, and the absence of pushing or pulling it, would be due to the unconventional fact that none of them, neither the atomic body nor the vacuum space, is dense and continuous. The rare networks of distant nuclear particles, which each of them represents, would thus easily penetrate and move through one another. The motion, when not very fast, would only cause vibrations of the nuclear particles, and wave motions in the networks, but not winds or currents.

The Anderson experiments, and scores of other experiments prove that in the vacuum space, there are bound electrons and positrons, far from one another. Known experimental data lead us to the conclusion that in and around our Solar System, the network formed by these particles is an electron-positron lattice, or epola, for short.

We have shown that the distance between the particles in the epola is about 50 times larger than their radii. Thus, the bound electrons and positrons occupy only a few thousandths of a percent of the volume of space. Hence, the vacuum space is quite empty, and does not resist the motion of atomic bodies, which represent even more empty networks of nuclei and electrons.

On the other hand, the quite empty vacuum space is far from being empty, because in one cubic millimeter of the vacuum space there is an amount of electron-positron pairs which can be represented by the number 6 10^{33}, or 6 followed by 33 zeros.

Hence, the combined binding energy of all pairs in one cubic millimeter of the vacuum space is a sixtillion joules, 27 quadrillion kilowatthours of energy, or 23 quintillions of calories. This *energy density* of binding is millions of times larger than in the binding of the outermost (valence) electrons in the atoms of a solid, but also a million times smaller than in the binding of the protons, neutrons, and other nuclear particles within the nuclei of atoms.

Adding or reducing a billion calories to or from the treasure of binding energies in a cubic millimeter of space means a gain or loss of only half a millionth of a percent of this treasure. These data clear the way to understand how the epola structure enables space to carry the observed tremendous fields of forces and radiation energies.

It should be clear that the ability of the vacuum space to carry such actions and energies can only be due to its physical structure. It cannot be and isn't due to mathematical, or philosophical derivations, no matter how widely (and expensively!) backed and cherished.

10.07. Reluctance to Accept the Epola Structure of Space

The reluctance to consider space a physical entity by people who have no financial, professional, or prestige interests in preserving the *status quo* in science, is based on our millennia old observations and experience that space does not resist the motions of atomic bodies, of the Earth and planets, and does not reduce their velocities.

This is simplest to explain by assuming that the vacuum space does not contain anything, that it is absolutely empty. Because we think of the Earth as being solid, continuous and impenetrable, it is clear and obvious to us, that if space were to contain any medium, which "obviously" must be continuous, too, then this medium would cause a resistance to the motion of Earth, and would have to be pushed, pulled, compressed and diluted by the motion. Thus, the motion of

Earth would have caused winds and currents in the medium.

When some resistance of space was later discovered and experimentally established, starting by 1919, along with many other explicitly physical properties of the vacuum space, the mathematics of empty space was already well developed. Hence, instead of bothering with the physics of observed phenomena, risking a rejection of their work, physicists can reach out for the "yellow pages" of this mathematics, pick up a ready or almost ready equation, and make all calculations, for which they would be paid, published and promoted.

Hence, hundreds of experimentally observed phenomena, which disclose the physical structure of space, like the bending of light, diffraction of particle beams, creation of waves around moving nuclear particles (de Broglie waves), the increase of the inertial resistance of nuclear particles to acceleration at high velocities, as well as inertia and gravitation, were not physically explained, and remain without a physical explanation till this very day.

These phenomena, and scores of other experimental facts, were interpreted mathematically, as resulting from such and such mathematical formula, under such and such mathematical operation or transform, as if mathematical expressions could cause and form physical phenomena. This misconception saved the postulated absolute emptiness of space, which remained and still is the pillar of "established science".

10.08. Lattice Structure of Sodium Chloride Crystals

The lattice structure of crystals can be revealed with the use of X-rays. The first to start the X-ray search of crystal structures were Max von Laue in Germany, and father and son, W.H.Bragg and W.L.Bragg, in England. In 1913, the Braggs published their results on the crystal structure of sodium chloride (rocksalt, or table salt). These results were validated by hundreds of followers, in numerous other measurements, at increasing accuracy levels.

The sodium chloride crystal is built of positively charged sodium ions, (symbol Na^+), and negative chlorine ions (symbol Cl^-). The positive charge of the sodium ion is equal to the charge of the positron, and the negative charge of the chlorine ion is equal to the charge of the electron.

The mass of the sodium ion is 23 atomic mass units (or amu), i.e., 23 times the mass of the proton or neutron, (and 23 times 1840 times the mass of the electron or positron). The mass of the chlorine ion is 35 amu. The radius of the sodium ion is 0.98 angstrom, and the radius of the chlorine ion is 1.8 angstrom.

The equilibrium distance between the centers of the nearest two sodium and chlorine ions in the crystal is the LATTICE CONSTANT of the crystal, and is equal to 2.8 angstrom. Hence, the distance between the closest spots of the two ions is altogether 0.02 angstroms. The binding energy of the two ions in the crystal is 8.0 electron-volts.

The crystal lattice, formed by the centers (the nuclei) of the sodium and chlorine ions in their equilibrium positions, is shown in Figure 1.

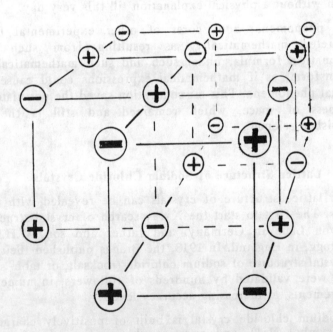

Figure 1. Unit Cubes of the Face Centered Cubic (fcc) Lattice of the Sodium Chloride Crystal

This is a cubic lattice (more exactly, a face-centered cubic, or fcc lattice), with the lattice constant l_o of 2.8 angstrom. The LATTICE POINTS, or lattice SITES are the equilibrium positions of the centers of the ions. Hence we may define the lattice constant as the distance between two nearest lattice points.

We introduce the term UNIT CUBE for a lattice cube of edge l_o. It can be seen on the Figure, that the 8 tips of the unit cube are occupied alternately by sodium and chlorine ions. Each of these ions "belongs" also to seven other unit cubes. Hence, only one eight (1/8) of each of these ions belongs to this unit cube.

Therefore, each unit cube, of volume l_o^3, contains altogether one ion. In other words, in the sodium chloride crystal, there is one ion per unit cube. We may generalize this, to say that *in a face-centered cubic lattice, there is one* LATTICE PARTICLE *per unit cube*.

10.09. The Sodium Chloride Lattice as an Analog to the Epola

The sodium chloride crystal lattice is the closest possible analog to the electron positron lattice for several reasons. First, the analog must be built of only two kinds of oppositely charged particles. The electric charge of the particle of the first kind must be equal to the charge of the positron, and the charge of the particle of the second kind must be equal to the charge of the electron.

Of all known existing stable lattices, this requirement is fit by 20 lattices of the ionic *alkali-halide* crystals, starting from the lightest, the lithium fluoride, then sodium chloride, etc., and ending with the heaviest, the cesium iodide crystal.

The second requirement is connected with the complexity of the binding between the atoms or ions in crystals. In addition to the electrostatic attraction of oppositely charged ions in the crystals, which yields the IONIC BINDING, there are always some other ways of binding involved, like the *covalent binding*, the *metallic binding*, etc., wich are well known to chemists.

However, the only attractive interaction in the binding of the epola is the electrostatic attraction between its electrons and positrons. We may say that the binding in the epola is "purely ionic", with no other kinds of binding involved, whatsoever. Therefore, out of the 20 alkali-halide candidates for an epola analog, we must choose the few ones with as little as possible of the unavoidable other components of binding, and with the highest percentage of ionic binding.

The third requirement is connected with the quadrillion times higher density of mass and of electric charge in the electrons and positrons than in the ions of the crystals, and with the hundred thousand times higher binding energy in the epola than in the alkali halide lattices. Therefore we expect that the amounts of these magnitudes per volume of a unit cube, i.e., especially the density of electric charge, and the density of binding energy, should be in the epola tremendously larger than in the ionic crystals, and in all atomic matter.

Hence, we have to choose, out of the few remaining alkali-halide candidates for an epola analog, the one with the highest density of charge and binding energy. And the winner is the sodium chloride crystal lattice. There are several other reasons, e.g., that the sodium chloride face-centered cubic structure is the most popular, repeated in 17 out of the 20 alkali halide crystals, that the sodium and the chlorine ions belong to the same *period* of the Periodic Table of Elements, and have therefore the same structure of their <u>inner</u> electron cores.

And last, but not least, the sodium chloride crystal structure is the best known, with the best, almost absolute fit between experimental, theoretical, and mathematical results. So that if we had not known about its fitting the requirements, we would have chosen the sodium chloride lattice anyway, just by the rule of thumb, that if you lose a dime in the dark, look for it at the nearest bright lights.

10.10. Calculation of the Electrostatic Attraction Energy in the Sodium Chloride and Alike fcc Lattices

Coulomb's Law allows us to calculate the electrostatic energy E of a "point charge", e.g., of an electron of charge -e,

$$-e = 1.6 \cdot 10^{-19} \text{ coulombs, (symbol C)},$$

in the field of another electron, or positron (charge +e), at a distance l between them. The energy is,

$$E = ke^2/l,$$

where k is a coefficient, the value of which depends on the choice of units. In SI units, the value of k is,

$$k = 8.85 \cdot 10^9 \text{ joule meter}/(\text{coulomb})^2,$$

$$\text{or } k = \sim 9 \cdot 10^9 \text{ J m}/\text{C}^2.$$

The energy of the electrostatic attraction of any ion of the sodium chloride crystal to the rest of the crystal was first calculated by E.Madelung in 1918, based on his "shell model" of the crystal. His calculation was made with the assumption that all ions of the crystal are in their equilibrium positions. Madelung found that this energy is 1.75 times larger than the electrostatic attractive energy of one ion in the field of another ion, outside the crystal, when the distance between the centers of the ions is equal to the lattice constant l_o of the crystal.

With Coulomb's Law, we thus obtain the formula for the electrostatic attractive energy E_{es} of an ion in a face-centered cubic crystal,

$$E_{es} = 1.75 \, ke^2/l_o.$$

The charge e of the ion is equal to the charge -e of the electron or +e of the positron, and is

$$e = 1.6 \cdot 10^{-19} \text{ coulombs}.$$

Substituting in the formula the values of k, e, and l_o,

$$l_o = 2.8 \cdot 10^{-10} \text{ m},$$

yields

$$E_{es} = 1.43 \cdot 10^{-18} \text{ joules}.$$

Replacing 1 joule by $6.25 \cdot 10^{18}$ electron·volts, we obtain

$$E_{es} = 8.89 \text{ eV}$$

Hence, the electrostatic attractive energy of an ion in the sodium chloride crystal is 8.89 electron·volt, by 0.89 eV larger than the known binding energy of two ions in this lattice.

It is a well known fact that electrostatic forces alone cannot keep a system of electric charges in equilibrium. Any system of electric charges, left to their electrostatic interaction only, must either collapse, or explode (**Earnshaw's Theorem**). Thus, if there were in the sodium chloride crystal the electrostatic interactions only between the ions, then the crystal would have collapsed, the positive sodium ions would just "fall" on the chlorine ions, and *vice versa*.

10.11. Calculation of the Short Range Repulsion Energy in the Sodium Chloride and Alike Lattices

For the stability of the lattices (also molecules), there must be a REPULSIVE interaction of non-electrostatic nature, which appears when the opposite ions are at a distance of less than two lattice constants apart. Hence, this is a SHORT-RANGE interaction. At a distance of 1.5 lattice constants, the repulsive interaction is very small, but grows very fast with decreasing distance between the ions.

When the distance between the ions becomes equal to the lattice constant, the repulsive force becomes equal to the attractive electrostatic force. As a result, when the ions are in their equilibrium positions at their lattice sites (in the lattice points), the sum of the attractive and repulsive <u>forces</u> acting on each ion is zero.

When the crystal is compressed, the distances between the ions are reduced below the lattice constant, the repulsive force is larger than the electrostatic attraction. Initially, this results in the observed <u>elastic</u> resistance of the crystal to compression. With further decrease in inter-ion distance, the repulsive force is skyrocketing.

The physical nature of the repulsive interaction was not disclosed in standard science, and remains unknown. However its formulae were approximated by various methods, to fit the experimental data. One of these formulae for the short-range repulsive energy $_{sr}E$, acting between an ion and the crystal, is,

$$_{sr}E = A \cdot e^{(-r\, l/l_0)}.$$

Here l_0 is the lattice constant, and l is the changing distance between the nearest neighbor ions in the crystal; r is a fitted *range parameter*. In the sodium chloride crystal, the value of the range parameter is 10. In other alkali-halide crystals, the range parameter has values from 7.2 to 12.5. The basis e of the power expression, is the basis of natural logarithms, $e=2.718...$.

In the formula, A is an *energy coefficient*, the value of which can be found from the condition that at $l=l_o$, the repulsive force becomes equal to the force of electrostatic attraction. In the sodium chloride lattice, the value of A is 2200 times the value of the electrostatic attractive energy E_{es},

$$A = 2200 \; E_{es}.$$

Hence the formula for the repulsive energy in the sodium chloride lattice is

$$_{sr}E = 2200 \; E_{es} \cdot e^{(-10 \; l/l_o)} \; .$$

In equilibrium, $l=l_o$, and $l/l_o = 1$. The formula for $_{sr}E$ is then,

$$_{sr}E = 2200 \; E_{es} \cdot e^{(-10)} = 2200 \; E_{es} \cdot 4.54 \cdot 10^{-5} = 0.1 \cdot E_{es} \; ,$$

thus

$$_{sr}E = 0.89 \; eV.$$

We see that in equilibrium, when the repulsive and attractive <u>forces</u> are <u>equal</u> to one another, the short range repulsive energy in the sodium chloride lattice is ten times smaller than the electrostatic attractive energy. As a rule, the range parameter r=10 is also the ratio of the electrostatic energy to the repulsive energy.

The total interaction energy E_{tot} in any lattice is the algebraic sum of the electrostatic and the repulsive energy. In equilibrium,

$$E_{tot} = E_{es} - {_{sr}E} = 8.89 \; eV - 0.89 \; eV = 8 \; eV \; .$$

Hence, the total interaction energy between an ion and the rest of the crystal is equal to the binding energy of an ion pair in the crystal.

10.12. The Lattice Structure of the Epola

In the sodium chloride crystal, all elements and data of the lattice structure were established both experimentally and theoretically to a perfect correspondence. However for the electron positron lattice, we have only the experimental value of the binding energy. Other data we have to calculate using the formulae for the sodium chloride lattice, which, according to the discussion in Section 8, owes to be the best analog for the epola structure.

First of all, we found that the epola lattice has to be face-centered cubic (fcc). We may also adopt the value r=10 of the range parameter in the sodium chloride crystal.

With the r=10 value, the repulsive energy $_{sr}E$ is ten times smaller than the electrostatic energy,

$$_{sr}E = 0.1\ E_{es}.$$

The total energy, E_{tot} is equal to the algebraic sum of these energies,

$$E_{tot} = E_{es} - {}_{sr}E,$$

so that

$$E_{tot} = E_{es} - 0.1\ E_{es} = 0.9\ E_{es},$$

and

$$E_{es} = E_{tot}/0.9.$$

In equilibrium, the total energy must yield the known binding energy of 511,000 eV per electron or positron in the epola. Therefore,

$$E_{es} = 511,000\ eV/0.9 = 568,000\ eV,$$

and

$$_{sr}E = 0.1\ E_{es} = 57,000\ eV.$$

With 1 eV=$1.6 \cdot 10^{-19}$ joule, the electrostatic energy, expressed in SI units, is

$$E_{es} = 568,000 \cdot 1.6 \cdot 10^{-19}\ \text{joule} = 9.1 \cdot 10^{-14}\ J.$$

The formula,

$$E_{es} = 1.75\ ke^2/l_o,$$

yields for the lattice constant l_o of the epola,

$$l_o = 1.75\ ke^2/E_{es}.$$

Substituting all the known values, we obtain

$$l_o = 1.75\ (9 \cdot 10^9\ Jm/C^2)\ (1.6 \cdot 10^{-19}\ C)^2/9.1 \cdot 10^{-14}\ J,$$

$$l_o = 4.4 \cdot 10^{-15}\ m = 4.4\ fm.$$

Hence, the lattice constant of the epola is 4.4 fermi (or femtometers, or 4.4 quadrillionths of a meter).

10.13. Limits for the Analogy Between the Epola and the Sodium Chloride Lattice

Calculations with values of r, extended beyond those which fit other alkali halide crystals, i.e., beyond 7.2 and 12.5 instead of 10, and with the use of the appropriate Madelung coefficients instead of 1.75, yield for the lattice constant in the epola, values between 4 and 4.8 femtometers. These values do not differ significantly from the 4.4 fm value, suggesting that our choice of the sodium chloride lattice as analog of the epola might be the right one for the calculation of various parameters of the lattice structure of the epola.

However the analogy cannot be drawn too far. Electrons and positrons are particles of nuclear matter, the presumptive "grains" or sub-particles of which (if any), are very densely packed. Even under "bombardment" with the now achievable trillion electron·volt energies (a billion times larger than their binding energy per pair) no electron or positron was ever destroyed or decomposed. Ions, on the other hand, are complicated planetary systems of nuclear particles, the nuclei and the electrons of which perform many rotations, vibrations, and pulsation. The energy of their decomposition (deionization or additional ionization) is only a few times larger than their binding energy in molecules or lattices. The interparticle distance of 4.4 fm in the epola is 50 times larger than the radii of the particles, which are below 0.1 fm. This is presented by the solid black circles in Figure 2. Hence, any bound particle "sees" its neighbor as a dense tiny sphere. Therefore, the bound epola particles interact as tiny charged spheres, or "point charges" *par excellence*.

In the sodium chloride lattice, as in many solids, the ions touch one another and may even overlap. This is shown by the large shadowed circles in Figure 2. The ions "see" the internal structure of each other, the rotations of the electrons on the internal elliptic orbits, and the rotations of the planes of these orbits. The supposed spherical shape of the ions, as well as their treatment as point charges, is here very problematic and very easily affected by internal and external forces.

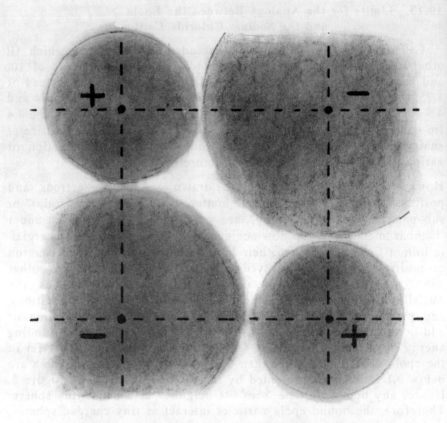

**Figure 2. Gates to Unit Cubes of the Epola
and of the Sodium Chloride fcc Lattices**

The black solid circles in the corners of the dashed square represent the positrons and electrons at the gate to an epola unit, in the scale of the 4.4 fermi lattice constant,- the side of the square. The free area of the gate is ~2000 times larger than the cross-section areas of the lattice particles, hence 99.9% of it is open for the passage of nuclear particles.

The shadowed circles show the ions of sodium and chlorine at the gate to a unit cube of the sodium chloride crystal in the scale of its 2.8 angstrom lattice constant. The radii of the ions are 0.98 and 1.8 angstrom. In this scale, the atomic nuclei would be invisible dots of a micron size. The open (white) area is 16% of the gate area, when thermal vibrations of ions can be disregarded.

Therefore, the properties of crystals depend strongly on the direction in the crystal, on the shapes of the surfaces, on the distances from them and from other boundaries, on the presence of defects in the structure, and so on. On the contrary, the epola is incomparably more uniform, homogeneous, and exhibits no significant directional effects, except, maybe, below the nanometer range.

We may therefore expect that physical processes, which occur both in the vacuum space (epola) and in atomic bodies, such as the electromagnetic and gravitational interactions, and the propagation of waves, will be revealed much easier in the epola than in the simplest (i.e., ionic) crystals. Also the unknown physical nature of the repulsive interaction in lattices should be easier to reveal in the epola.

CIRCLE FOUR

Chapter 11
FUNDAMENTAL PHYSICAL INTERACTIONS AND THE INTERACTION CARRYING SPACE

11.01. Interactions Presently Considered Fundamental

An interaction between physical entities is fundamental if it cannot be ascribed or deduced to other interactions. For example, two resting, static and stable particles, having electric charges of the same sign (i.e., both charges are positive, or both are negative) repel each other. This **electrostatic** repulsion cannot be ascribed or deduced to any other interaction. Similarly, the electrostatic attraction between two resting particles of opposite electric charges cannot be ascribed or deduced to any other physical (or whatever) interaction. Hence, the **electrostatic interaction** is fundamental.

Four physical interactions are presently considered fundamental. First is the **gravitational** interaction, which reveals itself in the mutual attraction of bodies, and is the stronger the larger their masses. It plays a decisive role (together with inertia) in the motions of atomic bodies on Earth, especially of large bodies, resulting in their observed heaviness, weight, and fall. In the cosmic scale, the gravitational interaction plays an almost exclusive role in holding the planets in their courses and controlling the motions of all bodies in the Solar System, again, together with inertia!.

Second is the **electromagnetic** interaction, combining the electrostatic interaction of electrically charged bodies with the magnetic interaction (attraction, repulsion, and turnovers) of bodies having magnetic moments. The electromagnetic interaction dominates in the atomic and molecular realms, and in the creation, propagation, and absorption of electromagnetic radiation.

The electromagnetic interaction is responsible for holding the atomic electrons and nuclei in atoms (again, together with inertia!). It is also responsible for all forms of binding atoms into molecules (molecular bonds), and of binding atoms, ions, and molecules into bodies of atomic matter. All other physical interactions in atomic matter, elastic and plastic, hydrostatic (buoyancy) and frictional, thermal and acoustic, can thus be deduced to the electromagnetic interaction, with the participation of gravitation (and inertia, of course). Furthermore, the interactions within living cells, and the signals they send to each other are electromagnetic, too.

The third fundamental interaction is **the strong nuclear interaction**. It coagulates (fuses) "elementary" nuclear particles into light nuclei, and light nuclei into heavier ones, when the distances between them are fractions of a fermi. The strong nuclear interaction is a hundred times stronger than the electrostatic interaction between charged nuclear particles (at a distance of 1 fm between them).

Fourth is the **weak nuclear interaction**, which brings together nuclear particles. It is orders of magnitude weaker than the strong nuclear interaction, but acts at distances up to slightly above a femtometer. Certain facts made some researchers state that the weak nuclear interaction can be deduced to the electromagnetic interaction. This would reduce the number of fundamental interactions to three. However others did not accept the reduction and still consider the weak nuclear interaction a fundamental one, with its name changed to "**electroweak interaction**". Both the strong and the weak nuclear interactions are responsible for the stability of nuclei and for nuclear transformations. These interactions cause all nuclear processes and the radioactivity of materials.

11.02. Ranges of Physical Interactions

The range of a physical interaction between bodies defines the distances between the bodies at which this interaction is in effect. For example, the electrostatic interaction between charged nuclear particles starts when the distance between the particles is at least one fermi (femtometer). This is the lower edge of the range of this interaction.

We have indications that the electrostatic interaction continues far beyond the Solar System, and we have no sign that it ceases at any particular distance. Hence, we consider the upper edge to be "infinity", or where the epola ends, "whichever comes first". Thus, the range of the electrostatic interaction is from 1 fermi to the borders or boundaries of or in the epola.

The range of the gravitational interaction has its upper edge also at such an infinity, while the lower edge is a hundred fermi, or a few, at least, epola lattice constants.

The strong nuclear interaction has a very narrow and small range from one to several tenths of a fermi. At its upper edge starts the weak nuclear interaction. The upper edge of the range of the weak nuclear interaction overlaps at a fermi the lower edge of the range of the electromagnetic interaction.

11.03. Interactions and Their Forces

Quite often physicists use the term "force" to denote an interaction. They speak and write of the gravitational force (also of the "fifth force"), the electromagnetic force, electroweak force, meaning not a particular force which acts on a particular body in the interaction, but having in mind the whole interaction. This causes additional confusion in the highly confused concept of force, and should be avoided, if not strongly objected.

It takes two to tango, and there must be at least two bodies for an interaction to occur. Then, each body acts with a force on the other. Hence, in each interaction there are at least two forces. "The gravitational force", strictly means a force due to gravitation, with which one body, say #1, acts on the body #2. "The gravitational interaction" contains also the gravitational force with which body #2 acts (or reacts) on the counterpart body #1.

According to Newton's Third Law, and to all our experience, the forces with which two bodies act on each other are always equal and opposite to one another. Denoting such forces by F_{12} (read "af one two") for the force with which body #1 acts on body #2, and F_{21}, for the force which body #2 exerts on body #1, we can write that

$$F_{12} = -F_{21}.$$

This is the shortest notation or formula of Newton's Third Law. The "minus" sign here indicates that the direction of one force is opposite to the direction of the other force.

11.04. The Interaction-Carrying Space and Fields;

Gravitational, electric, and magnetic interactions are actuated, correspondingly, by the masses, electric charges, and magnetic moments of the interacting bodies. However, unlike nuclear forces, which act only at an immediate (or almost immediate) contact between nuclear particles, these three interactions act also on distances, through the space between the interacting bodies. Clearly then, the space must be a participant in these interactions, and not only by its geometry, as Newton meant, but by its physical properties and structure.

It was established, 150 years after Newton, that if there occurs anywhere in space a redistribution of mass, charge, or magnetic moment, then the corresponding changes in the forces, acting on surrounding bodies, are not immediate. The changing interaction energies and forces propagate in space with the velocity of light c, which is c=300,000 km/s. They reach other bodies after a time t, which can be calculated by dividing the distance l to the bodies by the velocity of light. The formula for this calculation rule is $t=l/c$.

If the changes occurred 30 meters away from a body, then the variations in the interaction force and energy will reach the body after a time t, which is,

$$t = 30 \text{ m}/3 \cdot 10^8 \text{ m/s} = 10^{-7} \text{ s},$$

or 100 nanoseconds (100 billionths of a second). If they occurred on the Sun, they will reach us 500 seconds later.

Signals of changes, which occurred in other stars or galaxies, may reach us after years; tens, thousands, millions, and billions of years. During all these times, the changing interaction energies and forces propagate in space. Therefore, space has to be, <u>and is</u> recognized as the carrier of these energies and forces. But space is still not accepted as a physical entity of a profound physical structure, a structure which <u>enables</u> space to carry energies and forces.

The interaction-carrying space was granted the special name of "field". Thus, we have a gravitational field, an electric field, a magnetic field, and an electromagnetic field. It sounds special, but it does not add a thing to the understanding or to the trend to explore the physical structure of space, which is still considered Newton-wise and Einstein-wise empty, though accepted as the *official carrier* of all these fields, energies, and forces.

To make the empty space "able" to carry energies and forces, it was proclaimed deformable. If you ask how an emptiness can be deformed, you will be shown numerous mathematical derivations, which are proving, each in its way, that this is the way it is. If not convinced, you will be proclaimed an ignoramus, or just a fool, as were the unbelievers in H.C.Andersen's story about the new clothes of the king.

Numerous other physical properties were ascribed to the empty space, in addition to deformability, among them momentum, mass, and charge density, to enable the mathematical handling of phenomena in the accepted way. Each time new phenomena are discovered, additional physical features, imaginary mystical qualities, or tricks, have to be ascribed to the empty space to make mathematics work.

After the introduction of the epola and the derivation of its lattice structure, we can now make it clear that
    bodies act with their masses, charges, and magnetic moments
    directly on the bound particles of the epola
    in the space which they occupy and which surrounds them,
    deforming the space, i.e, deforming the epola.
This deformation propagates in the epola space with the velocity of light, becoming weaker with distance, as it spreads. When the deformation reaches an epola region, deformed by the masses, charges, and magnetic moments of other bodies, the particles of the epola exert on these bodies the corresponding interaction forces.

We do not see or perceive the surrounding air, except when it is dirty or windy. No wonder then, that we are unable to directly perceive or see the deformed space or fields. We do not even see the visible light; what we see are bodies which emit or reflect the light. In a ray of visible light, we see the tiny particles of dust, which reflect the light. But we do not see the ray, the light itself. What we detect in the deformed epola space are the interacting bodies. But it is the deformed epola space, which actually carries the gravitational, electric, and magnetic interactions, and exerts the interaction forces on the masses, charges, and magnetic moments on the bodies.

11.05. The Inertial Interaction of Two Atomic Bodies

Consider two atomic bodies, #1 and #2, moving on a horizontal path. Each body is electrically and magnetically neutral, and the gravitational and friction forces, acting on them, are balanced by supporting forces. The bodies move with constant velocities v_1 and v_2 relative to the surrounding space.

Suppose that the velocity v_1 of the first (leading) body is slightly smaller, and that they move along the same line or trajectory. Thus body #2 will at a certain moment contact body #1 from behind.

By inertia, body #2 tends to maintain its velocity v_2. However, starting from the moment of contact, this inertial trend is opposed by body #1, which, due to **its** inertia, tends to maintain the smaller velocity v_1. Therefore, body #2 starts pushing body #1 with a force F_{21} in the direction of motion, and body #1 reacts on body #2 with an equal and oppositely directed force F_{12}.

At the initial moment of contact, and at this moment only, both forces are purely and explicitly due to inertia, and cannot be deduced to any other physical action or phenomenon. The purely inertial interaction of these bodies is therefore a fundamental interaction, and the forces are the PROPER FUNDAMENTAL INERTIAL FORCES.

At the initial moment of contact, the atoms in the outer surface layers of the bodies are pushed inwardly, while the atoms of the inward layers, which continue their inertial motion, push to the outward, toward the other body. This creates a pressure wave in each body. The pressure wave and the related deformations of the bodies are deductible to the electromagnetic interaction, as also are the accompanying sounds, and the heating of the bodies. The interaction of the bodies is now mixed, not only inertial. The purely inertial interaction continues, with its forces decreasing, as the difference between the velocities of the two bodies decreases. The purely inertial interaction ceases when the velocities of the bodies stabilize.

The initial time-interval, during which the interaction is purely inertial, can be considered the time, during which the pressure wave moves from the atoms on the outer surfaces of the bodies to the next layer of atoms. This distance is around 3 angstrom ($3 \cdot 10^{-10}$ m), and the velocity of the pressure wave is the velocity of sound, which in solids is around 3 km/s. By the rule and formula $t=l/v$, we obtain

$$t = 3 \cdot 10^{-10} \text{ m} / 3000 \text{ m/s} = 10^{-13} \text{ s} = 100 \text{ fs}.$$

Hence, the time during which the whole interaction between the two atomic bodies is purely inertial, is around 100 femtoseconds (a tenth of a picosecond).

11.06. The Fundamental Character of the Inertial Interaction

The purely inertial interaction between atomic bodies is the "one and only" for only a short time of initial contact between bodies, and is very soon joined by interactions which are electromagnetic in nature (deformations, pressures and sounds, friction and heating, electrifying and magnetizing).

However this does not diminish the fundamental character of the inertial interaction. Fundamental nuclear interactions, too, last for a very short time, and are very soon replaced by forces and phenomena deductible to the electromagnetic interaction. The close range (immediate contact) at which the inertial interaction occurs, reminds the nuclear interactions, which, too, occur at immediate or almost immediate contact between nuclear particles. However the nuclear ranges are considerably smaller than the distance l_0 between the bound epola particles in our space, which is l_0=4.4 fm.

Hence, the fundamental nuclear forces act on nuclear particles directly, not through the bound particles in space. The bound particles become involved later, in secondary effects, e.g., by being either pushed out of their bonds (i.e., freed!), or forced into high-energy vibrations and wave motions (gamma rays).

To the contrary, the immediate contact range of the inertial interaction in atomic bodies is an angstrom, or the thickness of a monolayer (one layer) of atoms. This is equal to over twenty thousand lattice constants of the epola. Hence, the inertial interaction involves many thousands of monolayers of epola lattice units, and billions of bound epola particles.

11.07. Epola Deformation Caused By a Guest Particle

Consider that a nuclear particle entered an epola unit cube, of edge l_O, and volume l_O^3. The distance between the center of this particle and the 8 epola particles "sitting" in the tips of the unit cube is then equal to $0.87\ l_O$.

At this distance, the repulsive interaction of the "guest" particle with the 8 "host" particles is much stronger than the repulsion or attraction of epola particles at the l_O distance. Hence, the guest particle repels and pushes apart all eight host particles, electrons and positrons alike.

The repelled particles enter (diagonally) 8 out of the 26 nearest neighboring unit cubes. This results in particle displacements in all 26 unit cubes, and in the next neighboring layers of unit cubes. If the guest (or invader) does not leave, then the epola deformation spreads in all directions "to infinity", with the velocity of light.

It is important to note that at a distance of two epola lattice constants from the guest particle, the additional short range repulsion, caused by the guest directly, reduces to zero. The continuing spread of the deformation is carried by the electrostatic interaction between the displaced particles, and by the additional short range repulsions, introduced by the displacements of these particles themselves.

Hence, though the primary deformation of the host lattice unit is caused by the short range repulsive interaction, the spread of the deformation in the epola is caused mainly by the electrostatic interaction. We shall see that so it happens also when the primary deformations are gravitational or magnetic.

11.08. Epola Vibrations Caused by a Moving Guest Particle

Suppose now, that the guest nuclear particle moves in the epola with a velocity v, which is much smaller than the velocity of light c. Then every lattice unit, entered by the particle, expands. When the particle leaves the unit, the displaced epola particles move back toward their equilibrium positions, or lattice sites. Their motion is accelerated, with ever increasing speed, because of the lattice forces, acting towards the lattice points.

However, the bound particles cannot stop in the lattice sites, because of their inertia. They continue moving in the previous direction, now away from equilibrium. This motion is decelerated, until the particle stops at the maximum or amplitudinal displacement from equilibrium. Then the reversed accelerated motion begins toward equilibrium, and so on. Hence, the bound particles in an epola channel along the moving particle vibrate around their lattice sites.

The time during which the guest particle crosses a lattice region, causing expansion there, is equal to **one quarter of the vibrational period**. This is the time of one amplitudinal displacement. The next quarter of the vibration period is the time of return to equilibrium. The third quarter is the time of the amplitudinal displacement in the opposite direction, and the last quarter of each vibrational period is the time of return to equilibrium.

The duration of a quarter of the vibrational period, which is the crossing time of an expanding epola region, is the shorter the larger the velocity of the moving particle. Hence, the period of the epola vibrations, caused by the motion of a nuclear particle in it, is the shorter the faster the motion of the particle.

The frequency of vibrations is the inverse of the period. For example, if the period of vibrations is a thousandth of a second (a millisecond), then the frequency is 1000 vibrations per second, or 1000 herz (1000 Hz, or 1 kiloherz, 1kHz). Therefore,
*the frequency of epola vibrations,
enforced by the motion of a nuclear particle,
is directly proportional to the velocity of the particle.*

11.09. Accompanying Epola Wave Around a Moving Particle

The epola particles, vibrating in the channel along the moving nuclear guest particle, create an epola deformation wave, which propagates in the epola with the velocity of light. The frequency of this epola wave, which ACCOMPANIES the motion of the particle, is equal to the vibrational frequency of the epola particles in the wave. Hence, the frequency of the ACCOMPANYING WAVE is proportional to the velocity of the moving particle.

The wavelength of a wave is equal to the distance passed by the wave in a time equal to the period of the vibrations (and of the wave). The shorter the period, or the higher the frequency, the shorter the wavelength. The wavelength of the accompanying wave is therefore inversely proportional to the velocity of the moving particle. We have shown that the wavelength of the accompanying wave is also inversely proportional to the mass of the moving nuclear particle, and is equal to the "de Broglie wavelength" of the "waves of matter" introduced mathematically by L.V.de Broglie in 1924.

The accompanying wave propagates with the velocity of light, which is much larger than the velocity of the guest particle. Therefore, the wave has sufficient time to pre-form or to shape the epola in front of the particle, so that any epola unit, entered by the particle, is ahead of time properly expanded by the wave, to let the particle in, without efforts on its account. Hence, the once established motion of a particle, which cooperates with the accompanying wave and is co-operated by this wave, proceeds without any resistance of the epola.

The accompanying wave undergoes multiple reflections from the epola in front and behind the moving particle, as well as from the epola outside the channel along the path of the particle. Thus, the channel turns into a kind of "wave guide". Inside the epola channel along the moving particle, the vibrating epola particles form a succession of waves (a "wave train" or a "wave packet") which envelopes the moving particle and accompanies its motion.

The width or cross-section of the channel of the accompanying wave is always smaller than half the wavelength. The ratio of the width of the channel to the wavelength decreases at high velocities of the particle, the more the higher the velocity.

Hence, at high velocities, the width of the channel is reduced for two reasons. <u>First</u>, because the wavelength of the accompanying wave decreases with velocity. <u>Second</u>, because the width represents a smaller part of the wavelength. When the particle moves with a velocity close to the velocity of light, the width of the channel is zero. At such velocities, there is no accompanying wave.

For a free electron, moving with the velocity of an atomic outer orbital electron, i.e., $v=10^6$ m/s, the wavelength of the accompanying wave is about 1 angstrom, and the diameter of the channel is less than 0.1 angstrom. At an <u>electron</u> velocity of:

a jetfighter, v=1000 m/s, the wavelength is $\sim 10^{-7}$ m= 0.1 μm;
a racing car, v= 100 m/s, the wavelength is $\sim 10^{-6}$ m= 1 μm;
a horse cart, v= 10 m/s, the wavelength is $\sim 10^{-5}$ m=10 μm;
a pedestrian, v= 1 m/s, the wavelength is $\sim 10^{-4}$ m=0.1 mm.

At these velocities, the widths of the channels of the accompanying waves of the electron vary from tenths to a half of the wavelength.

11.10. The Physical Nature of Inertia

The widths of the channels of the accompanying waves represent the actual ranges within which the moving particles interact with the surrounding epola particles in directions <u>perpendicular</u> to the direction of motion. <u>Along</u> the direction of motion, the range of interaction spreads over tens and hundreds of wavelengths of the corresponding accompanying waves (or "waves of matter", in new physics). These measures are also the widths and lengths of the INERTIAL INTERACTION CHANNELS.

Two particles, the inertial interaction channels of which come into contact, or overlap, interact inertially with one another, altering the speeds or velocity directions of one another, just by inertia. Particles, for which the inertial interaction channels do not contact or overlap, are not interacting inertially. Hence, they do not INERTIALLY affect the velocities of each other (but may do so gravitationally, electrically, or magnetically).

We see that unlike the nuclear interactions, which initially do not involve the bound particles of space,
  the inertial interaction between two bodies
  results from the interaction of the particles of each body
  with the bound particles of space.
This is similar to the gravitational and electromagnetic interaction.

However, unlike these two interactions, which engage the bound particles in all available space, and spread upon them in all possible

directions, the inertial interaction
> involves only the bound particles inside the narrow channels of the accompanying waves around the moving particles.

We are now able to answer the almost 400 year old question of "**why there is inertia**". There is inertia, because
> around every nuclear particle of a moving body,
> there are billions and quintillions of epola particles,
> which vibrate at a frequency,
> proportional to the velocity of the particle.

As long as the velocity of the particle is constant, the vibrational frequency and energy of these tremendous amounts of epola particles remain constant.

Any change in the velocity of a nuclear particle requires a corresponding change in the vibrational frequency and energy of the tremendous amounts of epola particles, vibrating in the accompanying epola wave of the particle (or in the de Broglie wave train).

For example, starting the motion of a body relative to the surrounding epola, requires an energy supply to make the surrounding epola particles vibrate and create the accompanying waves around each nuclear particle of the body. The need to provide this energy is interpreted as due to inertia, to the inertial trend of the body to preserve its state of relative rest. The provided energy is then considered the energy which turned (after the reduction of losses on friction, etc.) into the kinetic energy of the body.

When one wants to stop the motion, he must provide outlets, able to absorb the energy of the accompanying wave (friction and heating of brakes, tires, and runway, raising ailerons and releasing parachute to increase air resistance, etc.). The need to provide such outlets is interpreted as due to inertia, to the inertial trend of the body to maintain its state of uniform motion, or to keep its velocity constant. The energy which must be taken away from the system, containing the moving body and its accompanying wave, is then considered the kinetic energy of the body.

We must agree that inertia is as much a "belonging" or feature of particles of nuclear matter, and of bodies consisting of such particles, as it is a property of the surrounding epola space.

Our derivation is qualitative, but it provides the ability to derive Newtons laws of motion, the formulae for the kinetic energy, the law of momentum conservation, etc. This ability is one of the indications, if not a proof, of the validity of the epola model of space.

11.11. Newton's Law of Gravitation

Based on his observations (the popular "falling apple" story depicts only one of plenty) and Laws of Motion, on Galileo's laws of free fall, and on Kepler's laws of planetary motions, Newton derived (in 1687) his Law of Gravitation, stating that:

two bodies, of masses m_1 and m_2, act on each other
with gravitational attraction forces F_{12} and F_{21},
which are proportional to the product $m_1 m_2$ of the masses,
and inversely proportional to the square of the distance l_{12}
between the centers of their masses.

This means, e.g., that if one of the masses is increased tenfold, then each of the forces increases tenfold. If the other mass is also increased, say five-fold, then altogether each force increases 50 times. If the distance is increased five-fold, then each of the forces becomes 5^2 or 25 times smaller. The short notation of the law is

$$F_{12} = -F_{21} = G \cdot m_1 \cdot m_2 / (l_{12})^2 .$$

The coefficient G is the "GRAVITATIONAL CONSTANT". It was calculated by Newton, and experimentally measured by H.Cavendish 111 years later (in 1798). The value of G depends on the choice of units. In SI units, the gravitational constant is,

$$G = 6.67 \cdot 10^{-11} \ Nm^2/kg^2 .$$

While the electromagnetic interaction can be either attractive or repulsive, the gravitational interaction is always attractive, and all attempts to find gravitational repulsion remained in mathematical equations, in papers only. The most recent attempt was the "fifth force" fever, at a cost of over 100 million dollars to missile-producing countries.

11.12. The "newton" Unit of Force

In the expression of the SI value of the gravitational constant, the letter N stands for the SI unit of force, which is 1 newton. The newton is a force, able to increase the velocity of a body of mass 1 kilogram, from zero to 1 meter per second, in one second.

In other words, the newton is a force, able to submit to a body of mass 1 kilogram the acceleration of a meter per second per second,

$$(1 \text{ m/s})/s = 1 \text{ m/s}^2 , \quad (1 \text{ meter per square second}).$$

The non-SI unit of force, the kilogram-force, is the weight of a body of a kilogram mass. In the free fall of a body on earth, the weight submits to the body an acceleration of 9.8 (almost 10) meters per square second. This is the "g" acceleration of free fall. Hence, the kilogram-force is 9.8 (or 10) newtons, and the newton is a tenth of a kilogram-force, or 102 gram-force.

Note. The explanation is much easier when the name of the kilogram unit of mass is changed to "lav", as suggested in Section 3.07.

11.13. Calculation of Gravitational Forces

The value of the gravitational constant,

$$G = 6.7 \cdot 10^{-11} \text{ Nm}^2/\text{kg}^2 ,$$

means that two bodies of masses 1 kg each, at a distance of 1 m between their centers, act on each other with a gravitational attractive force of $6.7 \cdot 10^{-11}$ N, or 67 trillionths of a newton, or 6.7 trillionths of a kilogram-force.

Let us calculate the gravitational attraction forces between two people, of mass 80 kg and 60 kg, when the distance between the centers of their masses is 10 m. We have

$m_1 = 80$ kg
$m_2 = 60$ kg
$l_{12} = 10$ m

$F_{12} = G \cdot m_1 m_2 / l_{12}^2 .$
$F_{12} = (6.7 \cdot 10^{-11} \text{ N})(80 \text{ kg} \cdot 60 \text{ kg}) / 100 \text{ m}^2 =$
$= \sim 3 \cdot 10^{-9} \text{ N} .$

This force, of 3 nanonewton, is good to lift 0.3 microgram, or a millimeter piece of human hair. Hence, people interested in mutual attraction should not count on the gravitation between them.

Let us now calculate the mass M of Earth. Knowing that Earth's radius is ~6370 km, and that at this distance from Earth's center, a body of mass 1 kg is attracted to earth by a force of 9.8 newtons, we can substitute these data to the equation of the Law of Gravitation. We have,

$$9.8 \text{ N} = (6.7 \cdot 10^{-11} \text{ Nm}^2/\text{kg}^2) (M \cdot 1\text{kg}) / (6{,}370{,}000 \text{ m})^2 .$$

We can now solve this equation for M, and obtain

$$M = 9.8 \text{ N} (6{,}370{,}000 \text{ m})^2 / (6.7 \cdot 10^{-11} \text{ Nm}^2/\text{kg}) =$$
$$= 9.8 \cdot 4 \cdot 10^{13} \text{ m}^2 / (6.7 \cdot 10^{-11} \text{ m}^2/\text{kg}) =$$
$$= 6 \cdot 10^{24} \text{ kg} ,$$

or a million quintillions of kilograms.

To "feel" the $6 \cdot 10^{24}$ kg mass of Earth, compare it with the mass of all inhabitants of our planet. Consider the number of people to be 4 billions, with a mass of 50 kg each. This yields for the mass of all inhabitants 200 billion kilograms. The mass of Earth is thus 30 trillion times larger. In other words, the mass of all inhabitants constitutes 33 quadrillionths of the mass of Earth.

Such a part of the 60 kilogram mass of a person, is 2 millionths of a milligram, or 2 nanograms. This is approximately the mass of a 2 micrometer "long" directly invisible piece of human hair. This means that losing all of us would mean to the mass of Mother Earth as much as a loss of this piece of hair means to the mass of a person.

11.14. On the Discovery of Neptune and Pluto

The Law of Gravitation enables the calculation of the masses of planets, as well as of any celestial body, if the distances to them and any effects of their gravitational action are known. Probably the greatest achievement of this law was the discovery of the planets Neptune (1849) and Pluto (1930), based on the observed disturbances in the motion of the planet Saturn. The discovery became possible when these disturbances were interpreted as resulting from gravitational interactions with unseen bodies.

Based on the Ptolemaic Model of the Solar System, which ruled in science for 1400 years, any disturbances in the motion of a planet are interpreted as due to the circling of the planet on an extra hypo-epicycle, the center of which circles on a hypocycle, the center of which circles on an epicycle, the center of which circles on the orbit. This "add a cycle" approach has nothing to do with reality, but can provide unlimited accuracy in calculating the positions of planets at any moment back and forth in time.

Such results were needed for calendars and horoscopes, and were paid for and funded by Kings and princes. The mathematically successful "add a cycle" approach kills any initiative to look for the physical reasons of the observed disturbances or phenomena in general. In the case of the disturbances in the motion of Saturn, it would have excluded the search for physical bodies behind the observables, hence it would have made the discovery of Neptune and Pluto impossible.

11.15. The Physical Nature of the Gravitational Interaction

In Section 7 we considered the deformation in the epola, caused by a free nuclear guest particle, which is resting inside a unit cube of the epola. Due to the strong short-range repulsive interaction, the guest particle repels the 8 epola particles positioned at the tips of the unit cube. The 8 displaced particles enter the surrounding unit cubes of the epola, displacing the particles there, too. The deformation spreads "ad infinitum" with the velocity of light.

Let us now consider that the guest particle is a neutron, of zero electric charge, and that at a certain distance of, say, a hundred lattice constants, there is another guest neutron in the epola. We could then show, that in the epola region between the two neutrons, the repulsive force, with which one neutron acts on each epola particle there, is opposed to the force, with which the second neutron repels this epola particle.

Therefore, the second neutron reduces the repulsive forces, exerted on the epola particles in this region by the first neutron, and vice versa. As a result, the repulsive forces, with which the deformed epola REACTS on the two neutrons, in the region between them, are equally reduced.

In the epola region <u>behind</u> the first neutron, the repulsive force, with which the second neutron repels each epola particle, acts in a direction, similar to the direction in which this epola particle is repelled by the first neutron. Equally, in the epola region behind the second neutron, the repulsive forces with which the first neutron repels the epola particles, act in directions similar to those, in which these epola particles are repelled by the second neutron.

As a result, the repulsion of the epola particles behind each neutron is increased, due to their repulsion also by the other neutron. Therefore, the repulsive force, with which the epola <u>behind</u> each neutron reacts, repelling it, is also increased.

The increased repulsion by the epola behind each neutron, pushes the two neutrons toward the region, where their repulsion by the epola is reduced, i.e., to the region <u>between</u> the two neutrons. Hence, the two neutrons are pushed by the deformed epola toward each other, with forces, which strongly increase as the distance between the neutrons decreases. These forces can be shown to be the forces of the gravitational interaction between the two neutrons, and explain the physical nature of the gravitational interaction.

We can now answer the 400 year old question of "**why there is gravity**". There is gravity because
    the deformed epola <u>between</u> any two bodies
    repels these bodies with forces which are weaker
    than the forces, with which the deformed epola <u>behind</u> each body repels the bodies <u>toward</u> each other.

This qualitative derivation enables us to derive the quantitative law of the gravitational interaction, leading to Newton's law. This possibility may serve as one of the indications, if not proofs, of the validity of the epola model for the physical structure of space.

In the empty region behind the first neutron, the repulsive force with which the second neutron repels each spoke particle acts in a direction similar to the direction in which the spoke particle is repelled by the first neutron. Equally, in the empty region behind the second neutron, the repulsive forces with which the first neutron repels the spoke particles, act in directions similar to those in which those spoke particles are repelled by the second neutron.

As a result, the repulsion of the spoke particles behind each neutron is increased due to their repulsion into two. On the other, at the same time, the repulsive force, with which the spoke particles each neutron repels, repelling it, is also increased.

The increased repulsion by the spokes behind each neutron pushes the two neutrons toward the region where their repulsion is weaker, i.e., reduced, i.e., in the region between the two neutrons. Hence, the two neutrons are pushed by the deformed spoke toward each other, with forces which sharply increase as the distance between them on decreases. These forces can be about 10⁴⁰ lb, the forces of the gravitational interaction between the two neutrons, and explain the potential nature of the gravitational interaction.

We can now answer the 400-year-old question of "Why there is gravity?" There is gravity, because:

The deformed spoke behind or any two bodies
(1) Is there a force with which they are repelled?
then the forces with which the deformed spoke behind each body repels the bodies toward each other.

This question's derivation enables us to derive the quantitative of the gravitational interaction, including the Newton's law. This possibility may serve as one of the indications, if not proofs, of the validity of the spoke model for the physical structure of space.

Chapter 12

THE ELECTROMAGNETIC INTERACTION AND THE ELECTRO-MAGNETO-GRAVITATIONAL (EMG) FIELD

12.01. Charge and Magnetic Moment Densities in Bodies

Let us consider a medium-light atom, e.g., of chlorine. It consists of 17 orbital electrons, and of a nucleus containing 17 protons and 18 neutrons. The mass of the electron is

$$m_e = 9.1 \cdot 10^{-31} \text{ kg},$$

and the mass of a proton or of a neutron is ~1840 times larger. Hence, the mass of the nucleus is 35·1840, or 64,400 times larger than the mass of the electron.

The charge of the electron is $-e = -1.6 \cdot 10^{-19}$ C (coulombs), so that the "charge per unit mass", e/m_e in the electron is

$$e/m_e = 1.6 \cdot 10^{-19} \text{ C} / 9.1 \cdot 10^{-31} \text{ kg} =$$
$$= 1.76 \cdot 10^{11} \text{ C/kg}.$$

This is a trillion times larger than the charge per unit mass in lightnings, and in our best condensers, which is ~0.1 C/kg (0.1 coulombs per kilogram of mass).

The charge of a proton is equal and opposite by sign (positive) to the charge of the electron. Hence the charge density in the proton is 1840 times smaller than in the electron. The charge density in the chlorine nucleus is (17/35·1840) or 3800 times smaller than in the electron, but is still gigantic.

The positive charge of the nucleus is equal and opposite to the negative charge of the orbital electrons, so that altogether the atom is electrically neutral. However, if any of the orbital electrons is deviated by external forces, then the electrical neutrality in certain spots of the atom is violated during certain, more or less measurable, time durations. Due to the gigantic charge densities in the atom, even the smallest deviations may create large electric and magnetic effects.

Large disturbances, e.g., in strong thermal collisions, may knock out an orbital electron, turning the atom into a monovalent positive ion, or they may push-in an additional electron, turning the atom into a negative ion. The charge per unit mass in a monovalent chlorine ion is 64,400 times smaller than in an electron, but is still gigantic.

The charge density (charge per unit volume) in a monovalent ion is ~10^{11} C/m^3. This is a quintillion times smaller than the charge density in the electron or positron. Still, the charge densities in ions are a trillion times larger than the charge densities in the most strongly charged atomic bodies.

In atomic bodies, excluding special condensers, or lightning *plasmas*, a charge density of a coulomb per m^3 cannot be achieved. For example, a sphere of radius 1 m (volume 4.2 m^3), at a voltage of ten million volts, would contain a charge of 1 millicoulomb only, and the charge density would be ~200 microcoulombs per cubic meter.

This is the charge of $1.2 \cdot 10^{15}$ electrons. In a cubic meter of a solid, there are ~$2 \cdot 10^{28}$ atoms. Therefore, the incredibly large charge of 200 microcoulombs per cubic meter of a solid, enforced by a voltage of 10 million volts, corresponds to only one excess electron (or excess ion) per 20 trillion atoms.

The charge density in the electron is 10^{32} times larger than the largest possible static charge densities in atomic bodies. These last are still thousands of times larger than the charge densities, which we encounter in everyday experience, as e.g., in the static electricity charges of cars, people, plastic rods, and other electrified bodies.

Quantitative characteristics of MAGNETIC MOMENTS require the introduction of a more complicated set of magnetic SI units, which we can do without for our purpose. We should mention without presenting calculations, that the densities of magnetic moments in strongly magnetized atomic bodies are also ~30 orders of magnitude smaller than the density of the intrinsic magnetic moment in the electron.

Hence, the observed "everyday" magnetism and static electricity of atomic bodies, are caused by tiniest deviations of the orbital electrons from their positions, relative to their orbits, and relative to one another, by deviations in the positions of their orbits, by the presence and distribution of some 'free' electrons and ions, and by the presence of a "handful" of excess electrons or ions.

12.02. The Spread of the Electromagnetic Interaction

The electromagnetic interaction is the most widespread; it is always here and there, everywhere wherever you stay or go. One may say that it is the Figaro of interactions.

There are several reasons for the interaction being widespread. <u>First</u> is the abundance in our space of bound electrons and positrons, which have extremely high densities of electric charge and magnetic moments. Any motion in space displaces these particles, which results in a more or less localized disturbance of electric and magnetic neutrality. Therefore, any interaction between particles is joined by, and ends up with electromagnetic interactions and radiation.

The second reason is the abundance of electric charges and magnetic moments of particles constituting the atoms, ions, molecules, and atomic bodies. Because of the tremendously high densities of electric charge and magnetic moments in these particles, the atoms, the molecules, and atomic bodies (including ourselves!) are actually gigantic reservoirs of electricity and magnetism.

Any interaction between bodies, starting with the most gentle sliding of a cloth on a plastic rod, or of a magnet on a needle, and up to the most catastrophic collisions, electrifies and magnetizes, while disarranging only a very small part of the charges and magnetic moments in the bodies.

The third reason for the widespreadness of the electromagnetic interaction is that it actually is a supermarket of interactions. Three of these, namely,
- the electrostatic interaction,
- the magnetic interaction of electric currents, and
- the interaction of the intrinsic magnetic moments
    of nuclear particles,

are each as fundamental, at least, as is the gravitational interaction. We shall therefore consider them separate fundamental interactions, and unite them, whenever appropriate.

12.03. The Electrostatic (Coulombic) Interaction

Influenced by Newtons Law of Gravitation, C.A.de Coulomb established in 1785 the law for the electrostatic interaction between point charges. "Point" charges, are material "points", carrying electric charges. These are mathematical imaginary presentations of electrically charged physical bodies, all sizes of which are much smaller than the distance to them or between them. Coulomb's Law states that

two point charges, q_1 and q_2,

resting in the surrounding space,

act on each other with forces F_{12} and F_{21}, which are proportional to the product $q_1 q_2$ of the charges, and inversely proportional to the square of the distance l_{12} between the point charges.

The short notation of the law is

$$F_{12} = -F_{21} = k \cdot q_1 \cdot q_2 / (l_{12})^2 .$$

The proportionality coefficient k in this equation expresses the interaction forces between unit charges at a unit distance between them. It was experimentally established by Coulomb. In recent SI units, the coefficient is,

$$k = 8.85 \cdot 10^9 \ Nm^2/C^2 .$$

This means that two POINT charges of 1 coulomb each would act on each other with forces of 9 billion newtons, if it were possible to keep them at a distance of one meter from one another. Mind you, 9 billion newtons would lift on earth 900 million kg, or three 300,000-ton large tanker ships. This points out how large the coulomb unit is in electrostatics.

However a coulomb isn't much in electric currents. For example, when a charge of a coulomb crosses a voltage of, say 110 volts, in a regular electric heater or other appliance, then the performed work is altogether 110 joules, or 0.026 of a "diet" calory.

Unlike gravitation, the electrostatic interaction is attractive between charges of opposite signs ("+" and "-"), and repulsive between charges of same sign ("+" and "+", or "-" and "-").

Another striking difference is the astonishing weakness of the gravitational interaction of nuclear particles and ions in comparison with their electrostatic interaction. This can be seen when comparing, e.g., the electrostatic attraction between an electron and a positron with their gravitational attraction, at one and the same distance l.

12.04. Comparison of Electrostatic and Gravitational Forces

For the electrostatic attraction between an electron and a positron we have, with Coulomb's Law,

$$F_{es} = k \cdot (-e) \cdot (+e)/l^2 = -k \cdot e^2/l^2.$$

The minus sign obtained here was accepted to indicate attraction. Hence we should insert this sign also in the expression for the gravitational attraction; thus

$$F_{gr} = -G \cdot m_e \cdot m_e/l^2 = -G \cdot m_e^2/l^2,$$

where m_e is the mass of the electron, as well as of the positron. Dividing the first equation by the second one, yields

$$F_{es}/F_{gr} = (k/G) \cdot (e/m_e)^2.$$

We have already calculated in Section 1 the ratio of the charge of the electron to its mass, which is

$$e/m_e = 1.76 \cdot 10^{11} \text{ C/kg}.$$

Let us now calculate the ratio k/G of the constants:

$$k/G = (9 \cdot 10^9 \text{ Nm}^2/\text{C}^2)/(6.7 \cdot 10^{-11} \text{ Nm}^2/\text{kg}^2) =$$
$$= 1.34 \cdot 10^{20} \text{ kg}^2/\text{C}^2.$$

Substituting these values, we obtain,

$$F_{es}/F_{gr} = (1.34 \cdot 10^{20} \text{ kg}^2/\text{C}^2) \cdot (3.1 \cdot 10^{22} \text{ C}^2/\text{kg}^2) =$$
$$= 4.15 \cdot 10^{42}.$$

The incredible result is that the electrostatic interaction forces between an electron and a positron are 42 orders of magnitude larger than the forces of their gravitational interaction!

Calculating the same for protons (and/or antiprotons), we recall that their charge to mass ratio is 1840 times smaller than in the electrons, and the square of this number is ~3.4 million. Thus, the ratio of the electrostatic interaction forces between protons, to their gravitational interaction forces, is 3.4 million times smaller than this ratio is for electrons; therefore,

$$(F_{es}/F_{gr})_{protons} = 4.15 \cdot 10^{42}/3.4 \cdot 10^6 = 1.2 \cdot 10^{36}.$$

Hence, the electrostatic interaction forces between two protons are 36 orders of magnitude larger than the gravitational forces.

The electrostatic interaction forces between two monovalent ions at a given distance between them are the same as between two electrons or two protons, at the same distance. However the gravitational forces between the ions are as many times larger, as the product of their masses is larger than the product of the masses of the protons or of the electrons.

For example, the gravitational interaction forces between the monovalent negative chlorine ion (Na^+, atomic mass 35) and the monovalent positive sodium ion (Cl^-, atomic mass 23), is (35·23) or 805 times larger than the gravitational interaction forces between two protons or neutrons at same distance. Therefore, the ratio of the electrostatic interaction forces between these ions, to the gravitational interaction forces between them, is 805 times smaller than this ratio is for a pair of protons; thus, the ratio is,

$$(F_{es}/F_{gr})_{NaCl} = 10^{33}.$$

The electrostatic forces, acting between such two ions, are still 10 quadrillion quintillions times, or 33 orders of magnitude larger than the gravitational forces, with which they act on each other.

The general rule is, therefore, that when one considers the electrostatic or magnetic forces acting on nuclear particles and on ions, he <u>must</u> completely disregard the gravitational forces, which are 40 and 30 orders of magnitude weaker.

We are used to feel the large gravitational forces exerted on us by earth. In contrast to these large gravitational forces, we experience the electrostatic forces as hardly detectable, able to lift some small pieces of paper or styrofoam. To achieve even this, one should take care that there is not too much vapors, ions, and breathing in the classroom. And suddenly these baby-orphan forces appear as super-super giants!

The seeming weakness of the electrostatic forces in our everyday experience is due to the overall electrical neutrality of atoms and bodies of atomic matter. The electric charges, which can usually be actuated or induced in atomic bodies, are due to very small displacements of charges, and to the very few electrons which can be added to or deduced from the tremendous amounts of electrons in the atomic bodies.

12.05. The Physical Nature of the Electrostatic Interaction

Consider a positively charged guest nuclear particle, resting in an epola unit cube. The distance between the center of this particle and each of the 8 epola host particles at the tips of the cube is 87% of the lattice constant l_0.

Therefore, the short range repulsive force, exerted by the guest particle on each of these host particles, is significantly larger than the electrostatic repulsive forces, with which the guest acts on each of the four host positrons, and than the electrostatic attractive forces, with which the guest acts on each of the four host electrons.

As a result, all eight host particles are pushed diagonally out, into 8 of the surrounding 26 unit cubes. However, the positrons are displaced slightly farther out than the electrons. The displaced particles cause the displacements of all particles of the 26 unit cubes, again with a slightly larger outward displacements of the positrons.

These displaced particles cause the displacements of the particles of the next layers of epola particles. At increasing distances from the guest particle, the layers become spherical. The short range repulsive force exerted by the guest particle is negligible already in the second layer, and is zero in the third one. In this layer, the positrons are pushed electrostatically outwards, while the electrons are pulled slightly inwards, toward the guest particle, which is positioned in the center of this and all the next spherical layers of charge.

Because of the outward displacement of the positrons and the inward displacement of the electrons, each layer of displaced epola particles consists of two concentric spheres: a larger sphere of positrons only, and a slightly smaller sphere of electrons only. As the distance increases, the split between the two spheres in each such "double layer of charge" decreases. In each layer, the excess (unscreened and unneutralized) charge is equal to the charge of the guest particle.

In other words, at any distance from the guest particle, the sphere of electrostatically displaced epola particles exhibits the extra charge of the guest particle. However the amount of excess charge *per unit area* of the sphere decreases with the increase of the area of the sphere. The area of the sphere is proportional to the square of the radius of the sphere, i.e., proportional to the square of the distance from the charged guest particle.

The amount of excess charge per unit area of the sphere yields the strength of the electrostatic field, or the force with which the field acts on a unit charge. From here it is a little step to derive the Coulomb law for the electrostatic interaction of electrical charges.

Our ability to derive the laws of the electrostatic interaction from the epola presentation of the electrostatic field is one of the many indications, if not a proof, of the validity of the epola structure of space.

12.06. The Magnetic Interaction of Electric Currents

Simple magnetic and electrostatic phenomena have been known for thousands of years. The first scientific book devoted to these phenomena was apparently William Gilbert's *"De Magnete"* (On Magnets), which appeared in 1600. Since ancient times, and till 1820, it was "common knowledge" that electricity and magnetism do not interact with one another and are physically distinct from one another, as well as from gravitation. However, attempts were made to unify their mathematical presentations in the form of Newton's law of gravitation.

The success of the mathematically analogous Coulomb's law for the electrostatic interactions, instigated many researchers to formulate a Coulomb- (or Newton-) like law for the magnetic interaction. All these formulations are either physically wrong, by assuming the existence of separate magnetic poles (monopoles), or too complicated and limited in applicability.

A turnabout occurred in 1820, when H.C.Oersted aimed to perform a classroom demonstration of the well known absence of interaction between electric charges and magnets. But when he placed a current-carrying wire above and along a magnetic compass needle, the needle turned almost perpendicular to the direction of the wire.

It then became clear that when electrically charged particles move, as in an electric current, or when the amounts of charges vary with time, then the particles interact also magnetically, as do magnetic bodies, i.e., bodies having magnetic moments. The magnetic interaction of bodies results in their attraction, repulsion, and turnovers (also flip-overs).

For example, two parallel wires, carrying electrical currents, attract each other, when the currents flow in the same direction (attraction of parallel currents); they repel each other, when the currents flow in opposite directions (repulsion of antiparallel currents). The magnetic moments play in the magnetic interaction a role, which reminds the role of electric charges in the electrostatic interaction.

The magnetic moments of current-carrying wires, loops, and coils, are fully due to the electric currents in them. In atoms, each orbital electron represents a circular current having quite a large magnetic moment. This is the *orbital magnetic moment* of the atomic electron.

Oersted's discovery initiated the development of a new branch of physics, named **electromagnetism**, especially thanks to the works of M.Faraday, A.M.Ampere, and to the electromagnetic theory created by J.C.Maxwell in 1865. Maxwell found also that the properties of static electricity and of the observed magnetism are connected with the velocity of light, that light is an electromagnetic phenomenon.

It became a new "common knowledge" that electric forces and the observed magnetic forces are just different aspects of the same phenomenon, and that all magnetic phenomena, including the whole magnetism of atoms, molecules, and magnets, are due to circular currents in them, and to the various alignments of these currents. Because of this interpretation, the magnetic interaction, fundamental by itself, was "unified" with the electrostatic interaction into one fundamental **electromagnetic interaction**.

12.07. Spin and Intrinsic Magnetic Moments of Nuclear Particles

In 1925, S.Goudsmit and G.E.Uhlenbeck discovered that each electron has a magnetic moment of its own, and actually represents a tiny little magnet. Because of the belief that any magnetism must be due to electric currents, the electron's own magnetic moment was interpreted as due to a rotation or spinning of the electron and of its charge around the axis of the particle. Such charge rotations would be equivalent to current loops. The magnetic moment of these current loops would then provide the magnetic moment of the electron. Hence the name "spin" for the electron's own magnetic moment.

Until the 1980's, the real radius of the electron was not known. The accepted value was 2.81 fermi, named "the classical electron radius", which was calculated from a speculation that the electrostatic energy of the electron must be equivalent to its mass (and don't ask why). The presently known <u>experimental</u> value of the radius of the electron is at least 30 times smaller.

However calculations even with the assumed 2.81 fm large radius showed that in order to yield the observed value of the spin magnetic moment of the electron, the velocities of its equatorial points would have to largely exceed the velocity of light. Hence, the spinning model was abandoned, leaving us with only the name "spin" for the since unexplained origin of this magnetic moment.

In studying the atoms, it was found that the spin of the orbital electrons plays a decisive role in atomic structure, together with the orbital magnetic moments of the electrons. This was followed by the discovery of the spin magnetic moments of the atomic nuclei, of protons, even of neutrons, the electric charge of which is zero. Thereafter, with the discovery of the thousands of particles of nuclear matter, it was found that each of them has its spin magnetic moment (except for some amalgamated particles of nullified spin).

For decades, the spin magnetic moments of nuclear particles was considered due to the alleged rotation (spinning) of their electrical charge around the particle axis. Such charge rotations would have been equivalent to current loops, too. Therefore, the whole magnetic moments of atoms, molecules, magnets, and of atomic bodies in general, were, and still are interpreted as due to circular currents of spinning and orbiting electrical charges, and to the various alignments of these currents.

However, the spin magnetic moments of nuclear particles are large, and can by no means be due to any possible or impossible rotations and pulsations of electrical charges in them. Moreover, the presence of a spin magnetic moment in the electrically neutral neutrons indicates that these moments do not result from current loops or rotations. Hence, the spin magnetic moments must be intrinsic to the particles, as parts of their **intrinsic magnetic moments**.

12.08. Magnetism of Currents and Intrinsic Magnetic Moments

We have shown that the magnetism of moving and varying electric charges results from their interaction with the charges and intrinsic magnetic moments of the bound electrons and positrons in our space. For example, a moving free electron simply sweeps the bound positrons of the epola slightly out of their equilibrium positions, with a turn toward the moving particle. The bound electrons of the epola are pushed slightly backwards from the moving free electron. This causes the observed magnetic interaction of currents and the deformation of the electromagnetic field (of the epola), explaining the close relation between electric and magnetic phenomena.

When free nuclear particles, whether charged or electrically neutral, interact with their intrinsic magnetic moments at distances much larger than the 4.4 fermi distance between the bound particles of the electromagnetic field, then this interaction involves tremendous amounts of the bound particles in-between.

With its intrinsic magnetic moment, each free particle actually interacts with the surrounding bound particles, deforming the field. The deformation of the field is spreading with the velocity c of the electromagnetic field, until it reaches the deformed region around the other free particle. Hence, the interacting particles do not "see" or "feel" each others intrinsic magnetic moments. The intrinsic magnetic moments of the interacting particles are "screened" and dissipated by the magnetism and the intrinsic magnetic moments of the bound particles in-between. What is left of the intrinsic magnetic moments of the free particles is their observed and measurable spin. Hence,
the spin magnetic moments of nuclear particles

with which they interact **within** the electromagnetic field, are the remnants of their intrinsic magnetic moments, which passed the screening of the epola particles in-between.

When two neutral nuclear particles approach each other to a distance of *4.4* fm and closer, then there are no bound particles between them. Hence, there is no screening or dissipation of the magnetic interaction in the region between the free particles. Thus, the two nuclear particles interact directly, with their pure, unscreened and unabridged intrinsic magnetic moments.

This is also a purely magnetic fundamental interaction, which is not due to the magnetism of electric currents, or to magnetic deformations of the electromagnetic field. On the contrary, the magnetism of electric currents, and the magnetic deformations of the electromagnetic field (epola) are due in part to the intrinsic magnetic moments of nuclear particles, free and bound, and can be deduced to them.

The unscreened interaction of the intrinsic magnetic moments of nuclear particles, at distances smaller than 4.4 fm , can be expected to be very strong. If the particles are also charged, then their magnetic interaction can be much stronger than their electrostatic interaction. Contrarily, at larger distances, the screened magnetic interaction between nuclear particles, i.e., the interaction of their spin magnetic moments, is known to be a hundred times weaker than the electrostatic interaction.

In our experience, the magnetic forces are much stronger than the electrostatic forces. While an electrified plastic rod can lift pieces of paper or styrofoam weighing a thousandth of the weight of the rod, magnetic forces exerted by a current loop in a light wire may lift a similar loop. Coils (solenoids) can do better, and a coil with an inserted core of a ferromagnetic material, or a so called electromagnet, can lift ferromagnetic bodies weighing significantly more than the electromagnet.

We should point out that the magnetization of ferromagnetic bodies, which is thousands of times stronger than in other atomic bodies, is to a large extent due to the alignment of the spins of their electrons and nuclei, i.e., to the screened intrinsic magnetic moments of their constituent particles, and not as much to the orbital magnetic moments, i.e., not as much to the magnetism of currents.

12.09. Stability Condition for a System of Bodies

We know very well, and also from everyday experience, that if a system of material bodies is left to gravitation alone (i.e., no more supports below, or hanging ropes above, and no inertial forces, as in the motion of planets), then the bodies cannot remain where they were, the system must collapse.

But when we consider a system of electric charges, left to electrostatic forces alone, the happenings are not so clear. Unlike masses, charges can be positive and negative; and unlike gravitation, which is attractive only, electrostatic forces can attract and repel.

Therefore, we often think, and so did many of our predecessors, that it might be possible to alternate positively and negatively charged bodies in linear arrays, on flat or bent surfaces, or in three-dimensional lattices, or whatever, in such a way, that this system, when left alone to the electrostatic attractive and repulsive forces only, would remain in equilibrium, or could be stable.

The creation of equilibrated electrostatic systems did occupy people, though not as much as the unfortunate "perpetuum mobile" did, until it was proven, experimentally and theoretically, (e.g., by Earnshaw's Theorem), that
>any system of electric charges,
>left alone to electrostatic forces only,
>cannot remain in equilibrium, or be stable;
>it must either collapse or explode.

Therefore,
>a system of electric charges can be stable,
>only if the electrostatic forces between the charges
>are balanced by forces of non-electrostatic nature.

This rule is quoted and used in some serious textbooks. Some other, otherwise serious texts, ignore the rule, for the sake of a quick explanation or derivation, and the ignorance reappears in classrooms, and among working physicists, too.

12.10. The Short Range Repulsive Interaction in Molecules and Lattices

The inevitable existence of the repulsive interaction between constituents of molecules became clear to chemists and physicists in the previous century. With its physical nature unknown, it was named "the short range repulsion", because it acts only in the distance range, shorter than twice the equilibrium distance l_o between the constituents of the molecule.

What was and is certain about the short range repulsion, in addition to its inevitability, is that when a molecule is in equilibrium, i.e.,
> when the distance l between the constituents is equal to l_o, $l=l_o$
> then the short range repulsion forces,
>> acting on the constituents of the molecule,
>> must be equal and opposed
> to the attractive forces between them.

Under this condition only is the sum of the attractive and repulsive forces yielding zero, and the molecule can be in stable equilibrium.

Another certainty is that at a distance $l=2l_o$, the short range repulsive forces are zero, so that they do not affect the attractive forces. From this distance down to $l=\sim1.3l_o$, the repulsion is weak, and has little effect on the attraction. The repulsion grows stronger and steeper as the distance decreases. At $l=\sim0.95\,l_o$, the repulsive forces can already be (e.g., in NaCl, the sodium chloride molecule) twice as strong as the electrostatic attractive forces.

It is well known, and formulated by Earnshaw's theorem, that electrostatic forces alone cannot keep a system of electric charges in equilibrium. Such a system, whether in the form of a particle, atom, molecule, or lattice, would either collapse or explode. Therefore, the short-range repulsion, necessary for the formation and stability of particle pairs, molecules, and lattices, cannot be electrostatic in nature, nor can it be due to any emanation or derivative of the electrostatic interaction.

The gravitational interaction between electrons and positrons, or between ions, is 40 or 30 orders of magnitude weaker than their electrostatic interaction. Thus, the short range repulsion cannot be gravitational in nature. It actually is the other way around, that the gravitation itself is due to the short range repulsion in the epola.

The range of the strong nuclear interaction is below a fermi, and the range of the weak nuclear (or "electroweak") interaction is around a fermi. Hence, nuclear interactions cannot constitute the nature of the short range repulsion, neither in molecules and solid lattices, where its range is around an angstrom, or 100,000 fermi, nor in the epola, where the upper edge of the repulsive interaction is 8 fm.

Therefore, of all known fundamental interactions, the short range repulsion in molecules and solid lattices can be magnetic, or inertial, or both. In the epola, the short range repulsion can only be magnetic.

12.11. Inertial and Magnetic Origins of the Short Range Repulsion

The inertia of the orbital electrons in the atoms is due to their interaction with the bound electrons and positrons of the epola around the orbits. This interaction creates the de Broglie accompanying space-waves, and is sufficient to prevent the orbital electrons from falling or collapsing on the nucleus. Thus, it equilibrates the electrostatic attraction of the orbital electron to the nucleus.

The electrostatic attraction between ions in the molecule is weaker than the attraction of the atomic electrons to the nucleus of the atom. Hence, the inertial interaction of the spacewaves of the outermost electrons of two atoms or ions, when deformed "on contact" at proximity, should also be able to provide sufficient repulsion to balance the electrostatic attraction between the constituent atoms in the molecule.

Each atom or ion of the molecule is a complicated system of many electrons, involved in very fast orbiting on non-simple orbits, the planes of which are also rotating and vibrating. All these motions create appreciable magnetic moments, in addition to the spin magnetic moments of the particles themselves.

When the distance between the ions exceeds $2l_o$, all this complicated space distribution of magnetic moments, as also of the electric charges of the particles, represents an "internal concern" in each of the ions. To the outside, they balance, (quite miraculously!) allowing the ion to appear as an analytically treatable point charge, having a certain non-zero or zero magnetic moment.

When distances between the centers of atoms or ions become smaller than $2l_o$, then the de Broglie spacewaves of their outermost orbital electrons begin to "touch". This means that the bound particles of space, residing in the overlapping volumes, have to participate simultaneously in these waves. This leads to an inertial resistance to the overlap, hence to a repulsion of the orbits, and to the repulsion of the atoms or ions to which the touching orbital spacewaves belong.

The "touching" of the orbital de Broglie spacewaves also evokes a repulsion of the orbital magnetic moments of the electrons, as well as of their spin magnetic moments. With decreasing distance between the ions, the repulsions strengthen; <u>first</u>, because of the increasing volumes of overlap of the de Broglie waves of the outermost orbits, <u>then</u>, due to the increasing number of outer orbits, which become involved in the overlap. <u>Finally</u>, inner shell orbiting electrons, with their accompanying de Broglie spacewaves, are forced into the interaction, too, resulting in the steep growth of the repulsion at distances below l_o. This corresponds to the behavior of the short range repulsive interaction in molecules and in solid lattices alike.

In the epola, the short range repulsion is due to the repulsion between the intrinsic magnetic moments of electrons and positrons at proximity. At a distance of two epola lattice constants ($2l_o$=8.8 fm), this interaction is still screened by two layers of epola particles, but it grows sharply with decreasing distance. At distances below the lattice constant, it is not screened, thus skyrocketing with further decrease in distance between the electron and positron. This corresponds to the behavior of the short range repulsive interaction in the epola.

12.12. The "Pauli Force" Deception

In teaching, the most popular explanation of the short range repulsion and stability of molecules and lattices is given as due to
"charge displacements in ions at proximity, which expose
the electrostatic repulsion between their bare nuclei,
or between their filled electron shells".
This is obviously deceiving, because it presents the short range repulsion as originating from the electrostatic repulsion of the bare nuclei.

More advanced texts base the explanation on the Puli Exclusion Principle, a basic one in quantum theory (see Section 16.12). It states that in a small volume, there cannot be two electrons having exactly all the same (quantized) energy values and spins (and no "why" asked nor answered!). The principle is then quoted as the originator of "Pauli forces", said to provide the stability of the molecule.

The "Pauli forces" are said to appear at distances l between the constituent atoms or ions, smaller than twice the equilibrium distance l_o. At this proximity, the outer electrons of the atoms become "forbidden" to one another by the Pauli principle (and no why!). This gives rise to the Pauli repulsive forces, which allegedly oppose the electrostatic attraction between the atoms or ions, and strongly increase with decreasing distance l.

We have shown, that the physics behind the Pauli principle is that two electrons, having the same spins and quantized energies (i.e., also identical velocities and trajectories, or orbits) must have identical de Broglie spacewaves. If the two electrons are confined to a proximity, at which their spacewaves overlap, then the interference of these identical waves unifies them into one wave (Section 16.12).

This means that the two electrons begin to move along the same orbit or trajectory. They are then exposed to their mutual electrostatic repulsion, until their accumulated electrostatic energy becomes sufficient to change the velocity value or direction of one of the electrons, or to flip over its spin magnetic moment. Each of these actions changes the spacewave of the electron, its orbit or trajectory, thus preventing any prolonged exposure of the electron to the electrostatic repulsion of the other electron.

The final result is exactly as postulated by the Pauli principle, i.e., that there won't be any such energetically identical electrons, but only in a long time average. During sufficiently short time intervals (picoseconds), electrons of forbidden energy states may exist.

Secondly, the physical interaction, making the Pauli principle work, is the electrostatic repulsion between the electrons in forbidden states, i.e., of electrons moving sufficiently long (picoseconds), with identical velocities, along the same trajectory. Hence, the Pauli forces are electrostatic in nature. As such, they cannot provide the short range repulsion, which <u>must be non-electrostatic</u> in order to prevent the collapse of the molecule.

Therefore, quoting the Pauli principle or Pauli forces as originators of the short range repulsion, is as deceiving as quoting for it the electrostatic repulsion of bared nuclei. The difference is that the latter "quick explanation" reveals the ignorance of the well established condition for the stability of any system of charges, and of the Earnshaw's theorem. The Pauli force "explanation" masks this ignorance, but reveals the tendency to replace physical reasoning by hide-outs behind unexplained "divine" principles and postulates, which in the long run is destructive to physics as a natural science.

12.13. Summary: Fundamental Interactions and the Electro-Magneto-Gravitational (EMG) Field in the Epola

We consider that our space contains electrons and positrons, bound to each other in a lattice (epola) at a distance of 4.4 femtometers from one another. These distances are ~50 times larger than the radii of these particles. Hence all known nuclear particles can move freely (if not too fast!) between the bound particles of space. The motion causes the bound particles to vibrate and create the de Broglie "waves of matter" or "spacewaves" around the moving particles. These are epola waves, accompanying the motion of the nuclear particles in the epola.

Atoms and atomic bodies consist of nuclear particles, which are twenty thousand times further apart than the bound particles in the epola space. Therefore, the epola space is also transparent for the undisturbed motion of atomic bodies, as it is to the motion of separate nuclei and electrons. The motion creates de Broglie accompanying waves around every nuclear particle of the moving atomic body.

(Note. Contrarily to what physics texts maintain and derive, there is no de Broglie wave of the whole atomic body; the spacewaves envelop each separate nuclear particle of the body, and do not sum up to one big brother wave.)

The epola structure of space allows us to explain the physical nature of inertia and gravitation. These interactions are initiated each by the masses of the nuclear particles of which the body consists. The interaction of the masses with the bound electrons and positrons is initially due to the short-range repulsive interaction, which deforms the surrounding epola. Therefrom, the gravitational deformations spread in the epola *ad infinitum*, while the inertial deformation is limited to a narrow wave-train around the moving particle.

En gros, these two epola deformations are electrically and magnetically neutral, thus not electromagnetic. However, the deformations displace bound electrons and positrons; hence, because of the unavoidable point-disturbances of electric and magnetic neutrality, gravitation and inertia are linked to the electromagnetic interaction.

Moreover, gravitation and inertia are both carried by the bound electrons and positrons, i.e., by the electromagnetic field, as is the electromagnetic interaction. Hence, the two may lose their licence for being fundamental. This would bring all out-of-nuclei interactions to one fundamental interaction.

Whether one accedes or not to this unification, he must agree that the gravitational field, the electric field, and the magnetic field, are one and the same physical entity, which he may call the Electro-Magneto-Gravitational field, or the **EMG field**, for short, as we do.

Some interactions between atomic bodies are considered to occur at immediate contact, like friction and deformations. In these interactions, the distances between the orbital electrons of the atoms, or between the atomic nuclei, are in the order of an angstrom. This is 20,000 times larger than the 4.4 fermi distances between the bound electrons and positrons of the EMG field.

Thus, the atoms of atomic bodies, interacting "at immediate contact" are separated by some 20,000 layers of bound electrons and positrons. Hence, the nuclear particles, constituting the atoms of the interacting bodies, actually interact with the hundreds of layers of the adjacent bound particles. Thus, the interaction between the bodies "at immediate contact" is carried by the bound particles in-between.

Therefore, due to the micro-structure of space, there is no <u>principal</u> difference between interactions of atomic bodies on contact, i.e., at a distance of an angstrom, or at any larger distance, because all interactions between atomic bodies are anyhow carried by the EMG field.

Only free nuclear particles can approach each other to distances much smaller than an angstrom, thus reducing the number of the bound particles in-between. When free nuclear particles are at distances of 4.4 fermi or less, they interact without bound particles between them. They are then interacting with their pure unscreened intrinsic magnetic moments and with their electric charges.

However, the free particles are still under the influence of the surrounding bound particles, i.e., of the "big brother" EMG field. At a distance of, or below, 1 fermi, the nuclear forces take over, subjecting the particles to collapse and coagulation. But even then the EMG field plays a role in the creation, behavior, and decay of the coagulant.

It may seem contradictory that we provide arguments for the unification of gravitation and inertia with the established electromagnetic interaction, while on the other end we resolve this interaction into its three components: the electrostatic, the intrinsic magnetic, and the magnetic interaction of electric currents. We do this for the sake of consistency: these three interaction are not less unique and fundamental than those two. Thus, we should either have one consisting of five, or five equally fundamental and interconnected interactions.

The electron-positron structure of space allows us to fully understand the physical nature of these five interactions, i.e., to explain WHY bodies have inertia, why they gravitate, why parallel currents attract each other, and so on. However, this does not reveal the physical nature of mass, electric charge, and intrinsic magnetic moment. These are intrinsic properties of nuclear particles, which we can measure, because of the named interactions, but we do not know what they are and why they are as they are. *Ignoramus*.

We may say in self-defense that our topic is the structure of space, not the physical nature of the intrinsic properties of nuclear particles, and not of the nuclear interactions. These topics belong to the "Nuclear Particles" and "Particles and Nuclei" branches of physics. To the best of our knowledge, these branches are not yet ready to deal with such subjects. Hopefully, consideration of the electron-positron structure of the EMG field may be of some help and stimulation in clearing the physical nature of nuclear particles, when the respective physicists shall become interested in such a clearing.

CIRCLE   FIVE

Chapter 13
BULK DEFORMATION WAVES IN BODIES

13.01. The Thermal Motion of Constituent Particles in Solids

In solids, as also in liquids, the constituent molecules, atoms, or ions, are strongly bound to one another. The thermal motion in solids and liquids consists therefore of vibrations only of the constituents around their equilibrium positions in the body, e.g., around the lattice points (or *lattice sites*) in crystals.

The amplitude of the thermal vibrations is much smaller than the average distance between the constituents. Displacements or travel beyond the average distance, which is common in gases, can happen in a solid or liquid, only if the constituent gains an energy larger than its binding energy, and is freed out of the bonds with its neighbors. Though the freed particle may remain in the body, it is not a constituent but an "ex", until captured back into the bonds. Atoms and molecules, seldom ions, can also be freed out of the body, and evaporated from its surface.

The thermal vibration energies of the constituents of a body differ very much from one another. The higher the temperature of the body, i.e., the richer it is in energy, the larger the differences in energy between the "rich" and "poor" constituents. The energy distribution is the closer to "equality" the lower the temperature. A full equality could be reached at the unreachable *absolute zero* temperature, at which the thermal energy of each constituent would be zero.

The absolute zero temperature was calculated to be about -273.15 degrees Celsius (°C). The temperature measured from this absolute zero in units of **kelvin** is the **absolute temperature**. The kelvin unit, symbol K, is equal to a degree Celsius. However the absolute temperature "contains" 273.15 more kelvins than degrees Celsius. For example, the temperature of 0 °C, at which water freezes, is 273.15 K.

It is not yet possible to determine the energy of a particular constituent at a particular moment. But from the measurements of the "specific heats", i.e., of the amounts of thermal energy, needed to raise the absolute temperature of a body by 1 kelvin, and knowing the number of constituents in the body, it is possible to calculate the mean thermal energy of a constituent.

The thermal motions in a body are perfectly random. At any particular moment, each constituent moves in a certain direction with a certain speed, i.e., has a certain velocity "*vector*". However other constituents move each in its own direction, each with a velocity vector of its own.

Note. The sum of *vector* magnitudes, like forces, accelerations, velocities, depends on their directions. Thus, the sum of two forces, say 10 newtons each, acting simultaneously on a *material point*, can have any value between zero and 20 newtons, depending on the angle between the directions of the forces.

As a result of the perfect randomness of the motions of the constituents, the sum of their velocities is zero. For example, let one particle momentarily have a velocity of 100 m/s, directed along the y-axis to the left from the point of origin of a chosen coordinate system. Among the sixtillions of particles around, there will certainly be a similar particle, which at the same moment moves with a velocity of 100 m/s in the opposite direction. Hence, the sum of the velocity vectors of these two particles is zero.

Whatever the velocity vector of any constituent particle of the body, there always is, among the sixtillions of particles in the body, a particle having an opposite velocity vector, so that the sum of the two velocity vectors is zero. Hence, the sum of the thermal velocity vectors of all constituent particles of a body is zero at all times. Therefore,
> the thermal motion of constituent particles
> cannot make the body move,
> nor can it change the motion of the body.

This is so, as long as the particles <u>are</u> constituents; when they move out of the body, by evaporation, explosions, or jet streams, they are no longer constituents of the body, and the rule does not apply.

13.02. Mean Energies of Thermal Motion

L.E. Boltzmann found (by 1870) that the average thermal energy of a constituent particle in a body is proportional to the absolute temperature T of the body. For example, at an absolute temperature T=300 kelvin, which is 27° Celsius, or 80°F, the mean thermal energy of a constituent particle of a body is around 0.04 electron·volts. At a two times higher absolute temperature of 600 K, which is 327°C, the mean thermal energy of a particle is two times higher, and is around 0.08 eV.

In the sodium chloride crystal, the binding energy of an ion pair is 8 eV, but the energy of freeing ONE of the ions is above 5 eV. The mean thermal energy per ion at 300 K is ~0.04 eV, so that the probability of an ion accumulating the freeing energy from the thermal motions of its neighbors is very close to zero. Therefore, thermal freeing of an ion within a pure and perfect crystal, which does not contain mobile impurities and imperfections, is at this temperature almost impossible.

13.03. Conditions for the Formation of Bulk Deformation Waves

Due to their randomness, the thermal vibrations of the constituents of a body in thermal equilibrium cannot create steady, long-lasting waves in it. At every instant, there are certain groups of constituents in the body, which would align their thermal vibrations coherently to form seeds of waves of certain frequencies. These are outbursts of "thermally excited" waves. Most of them will soon have their energy "randomized" by the thermal motions of the surrounding particles, and will not survive a nanosecond.

Because the wave velocity in solids is around 3000 m/s, the surviving waves will in microseconds reach a boundary, an edge, face, or surface layer of the body. Some waves may be absorbed there, but even the reflected ones do not last much longer. Trillions of surrounding randomly vibrating constituents would have already disordered their seeds, so that the waves have no energy support to replenish what they lose on randomly vibrating particles of the body. Hence, the lifetimes of the thermally excited waves are quite short.

To start a wave motion, an adequate number of constituents of the body must gain a sufficient amount of vibrational energy from an outside source. Starting to propagate from this seed, the wave reaches consecutive spots in the body. Each spot, or any group of constituents, reached by the wave, must vibrate according to the commands, or "phases", of this wave, independent of the thermal and other vibrations of the spot. The spot can then be considered as a center, from which a new wave propagates, not only forwards, but also backwards, and in other possible directions.

The additional condition for the establishment of a wave is that the backward or reflected waves from each spot must reach the *"veteran"* vibrating spots of the body, as well as the seed of the wave motion, with *"commands"*, or phases, identical to the phases of these spots at this instant. The interference of the reflected wave with the propagating waves from these spots must not be destructive.

A bulk deformation wave is not easy to depict or imagine. To us, a wave is what we see on an elastic string, or on the water. These are nearly *"transverse waves"*, in which the displacements of the vibrating spots are perpendicular to the direction of propagation. *Longitudinal waves* in these media, i.e., waves in which the particles of the media are vibrating <u>along</u> the directions of wave propagation, are already hard to visualize.

Now try to visualize a bulk deformation wave, which is neither transverse nor longitudinal. In the bulk wave, the strongly bound particles vibrate in all possible directions, in addition to their thermal and other vibrations, with amplitudes much smaller than the sizes of the particles, or than the distances between them.

13.04. Half-Wave Deformation Clusters of Bulk Waves

We consider the bulk deformation wave as consisting of half-wave deformation clusters. On one wavelength of the wave, there are two half-wave deformation clusters. If one half-wave cluster contains momentarily a larger number of bound particles of the body than in equilibrium, then in the next, or previous half-wave cluster, the number of particles is smaller than in equilibrium. The propagation of the wave can thus be thought of as the propagation of alternating half-wave regions of increased and reduced particle density, or as the propagation of CHANGES IN PARTICLE DENSITY.

However, it has to be clear that the particles in the wave are molecules, atoms, or ions, which are bound in the body, and can move less than a nearest neighbor distance l_o. All they can do is vibrate around their equilibrium position in a certain direction, increasing the density of particles in this direction, and reducing the particle density in the backward direction. Half a period later, they move backwards, increasing the particle density in this direction, and reducing the particle density in the forward direction.

Increasing the particle density, i.e., reducing the inter-particle distance, leads to an increase in the total interaction energy of the particles, i.e., to an increase in the sum of their short-range repulsive energy and their attractive electrostatic energy. Reducing the particle density, i.e., increasing the inter-particle distance, leads to a reduction in the total interaction energy. In equilibrium, the total interaction energy is the binding energy $_bE$. Thus, the binding energy is also the average of the increased and reduced total interaction energy in the wave.

Hence, in the half-wave deformation cluster with a momentarily increased number of particles, the density of the interaction energy is increased. In the half-wave cluster with the reduced number of particles, the interaction energy density is reduced. The propagation of a bulk deformation wave in a body is therefore the propagation of half-wave clusters of not only increased and reduced particle density, but also of increased and reduced interaction energy density.

It is important to remember that the particles of the body in any half-wave deformation cluster do not move with the wave, they remain around their equilibrium positions or lattice sites. Moving with the wave are changes in the interaction forces, causing increases and decreases in particle densities, as well as changes in interaction energies. They result in a net energy transfer or "flow" of energy with the wave.

In other words, the bulk deformation wave
    transfers the changes of the particle density in the medium, and
    the changes in the density of the interaction energy
        of the constituent particles of the medium.

The average between the increased and reduced **total** interaction energy of a particle in the wave is the binding energy of the particle in the equilibrium position.

13.05. Excess Particles and Energies in Bulk Deformation Waves

Consider a bulk deformation wave, propagating in a large sodium chloride crystal. Let the wavelength be 2800 angstrom, thousand times larger than the lattice constant of the crystal, which is

$$l_o = 2.8 \text{ angstrom} .$$

Each half-wave deformation cluster can then be considered spherical, of radius R=700 angstrom, and volume V=1.4 billion cubic angstroms.

In the lattice, the unit cube volume l_o^3 is 22 cubic angstroms, and contains **one** lattice particle. Therefore, the equilibrium number N of particles (ions) in the volume of a half-wave cluster can be obtained by dividing the volume V of the cluster by the unit cube volume. Hence, this half-wave cluster contains in equilibrium 64 million host particles.

The number N of particles in the half-wave cluster is proportional to the cube of the radius, hence also to the cube of the wavelength. Denoting the wavelength by w, we can write this rule as

$$N = k_1 w^3 ,$$

where k_1 is a proportionality factor, not defined by the rule.

The average binding energy per particle, or per l_o^3 unit cube, is $_bE$=4 eV. The half wave, which contains an excess number n of ions, thus also contains an excess binding energy E, of $n·$4 eV. The other half-wave has the number n of ions missing, and a missing binding energy of E=$n·$4 eV. Therefore, the propagating wave is transferring an excess (or missing) energy of plus or minus E, E=$n·$4 electron·volt.

The excess particles are bound in the crystal, as are all its particles. Therefore, their displacements in vibrations between the two half-wave clusters are always smaller than l_o. This means that during the wave propagation, the excess particles remain in a surface layer of the one or the other half-wave cluster, and that the thickness of the surface layer is less than l_o, in the order of the size of an ion. This surface layer is therefore a so called *monolayer* of ions.

By shifting towards a given cluster, the monolayer of ions causes all ions in the cluster to shift toward each other, increasing the particle density in the cluster above the equilibrium value. When the monolayer shifts toward another cluster, all ions in the given cluster shift apart from each other, reducing the particle density in the cluster, short of its equilibrium value.

Therefore, the number n of the excess (or missing) ions in a half-wave deformation cluster is actually the number of ions in the monolayer on the surface between the clusters. Hence, n is proportional to the surface area A of the half-wave cluster. The area A of the cluster is proportional to the square of the radius of the cluster, thus to the square of the wavelength, to w^2.

Hence, the excess energy E of a half-wave cluster is also proportional to the square of the wavelength, to w^2. We can express this rule in a short notation,

$$E = k_2 w^2 ,$$

where k_2 is another undefined proportionality factor.

13.06. Energy Transfer by Wave Clusters and by Single Particles, or in Quanta

The energy E, transferred by the half-wave deformation cluster, is proportional to the square of the wavelength. Thus a target, large enough to receive the energy E of the <u>whole</u> half-wave cluster, absorbs an energy proportional to the square of the wavelength. Targets smaller than the cross-section area of the half-wave cluster, absorb a correspondingly smaller part of its energy, and of the energy of the wave. The smaller the target or detector, the smaller the part of the wave energy which it absorbs.

However, the wave energy absorbed by a target, or detector, cannot be smaller than the energy E_p, transferred in the wave by a single constituent (ion, atom, molecule) of the body. A target of size, comparable to the sizes of ions, may not absorb the wave energy at all, or may absorb the energy transferred in the wave by one ion, two ions, or any appropriate integral number of constituents. It cannot absorb (and there is no such thing as) the energy transferred in the wave by half an ion, or seven quarters of a vibrating atom.

Hence,
> the energy transferred in a wave motion
> by the constituent particles of a solid or liquid body
> is quantized; the quantum of this energy is
> the wave energy E_p, transferred in the wave
> by a single constituent of the body.

To calculate the energy E_p per constituent particle in the wave, we divide the energy E of a half-wave cluster by the number N of particles in it. Therefore,

$$E_p = E/N = k_2 \, w^2/k_1 \, w^3 = k_3/w \, .$$

Here $k_3 = k_2/k_1$, is another undefined proportionality factor. We see that the per-particle energy of a half-wave cluster, thus also the per-particle energy of a bulk deformation wave in a solid or liquid medium, is inversely proportional to the wavelength. This per-particle energy is actually the energy quantum of the wave.

13.07. Physical Derivation of Planck's Postulated Law

The wavelength of a wave can be calculated as the quotient of the propagation velocity v_s of the wave by the wave frequency f,

$$w = v_s/f \, .$$

In the formula, which we derived in Section 6 for the quantum energy E_p of a wave,

$$E_p = E/N = k_3/w \, ,$$

we thus replace the wavelength w by v_s/f, and obtain

$$E_p = k_3/w = (k_3/v_s) \, f \, ;$$

hence,

the energy quantum of the wave is proportional to the frequency.

Replacing the proportionality factor k_3/v_s by another yet undefined proportionality factor h',

$$h' = k_3/v_s = (k_2/k_1)/v_s \, ,$$

we obtain

$$E_p = h' f \, .$$

Hence,
when a bulk deformation wave propagates in a medium, then
the energy E_p, transferred in the bulk wave
per constituent particle of the medium, carrying the wave,
is the quantum of the wave energy, and
is proportional to the frequency of the wave motion.

The proportionality of the quantum energy of light waves to their frequency was first introduced in 1900 by Max Planck as a postulate, without an explanation, derivation or proof. It is still unexplained in quantum mechanics till this very day, and the proof of it is in that "it works". And because it works so well, the postulate is related to as Planck's Law. The accepted formula of the law is

$$E_p = h f \, ,$$

and reads that the energy of the photon, or quantum of light, is proportional to the frequency of the light wave. Here the proportionality factor h is Planck's Constant. Its value was laboriously derived by Planck, and is

$$h = 6.63 \cdot 10^{-34} \text{ joules/hertz, or joules·second.}$$

When our derived formula for the quantum energy E_p of a wave motion is applied to light or other electromagnetic waves, then the value of the proportionality coefficcient h' is identical to Planck's constant h. The coefficient actually shows the quantum energy of a wave of unit frequency.

13.08. The Phonon-Photon Analogy

The phonon concept was introduced by the quantum theory of solids in the 1930s as an analog to the photon. Just as the photon was initiated as the quantum of electromagnetic radiation energy, the phonon was launched as the quantum of acoustic energy. And as the photon energy was postulated by Planck to be proportional to the frequency f of its electromagnetic wave, so was the phonon energy postulated to be proportional to the frequency f of its acoustic wave.

However, it soon turned out that there are no phonons in gases, and not in the free air, in which most of our acoustics occurs. It became a quietly agreed fact that the quantization of acoustic energy, occurring in solids and liquids only, must be due to the order and binding of constituents in liquids and solids, particularly to the long-range order in crystals. The phonon was therefore reduced in rank to a quantum of lattice vibrations and waves in "condensed matter" only.

Reversing the phonon-photon analogy, one may consider that
    if the quantization of acoustic energy
    is due to the lattice structure of solids, or
    to the lattice-like "short range order" in liquids,
then the quantization of the electromagnetic radiation energy might be similarly due to a definite lattice structure of the electromagnetic field.

And if the "divinely" postulated proportionality of the phonon energy to the wave frequency can be derived by mortals so easily, as in Sections 6 and 7, then it might be possible to achieve the same for photons. Hence, the still unexplained Planck's postulate could be replaced by a physical derivation of the per-particle wave energy, based on the proportionality of the wave energy to the square of the wavelength.

Being aware of the danger of such physical approaches to the domineering of quantum theory, the theorists claim that the quantization of the wave energy is not a result of the physical structure of bodies, but of the omnipotence of quantum operators and equations. Hence, they claim that there is no physical basis for the phonon-photon analogy. Wherever observed, such an analogy is to them a result of the "unity" and "universality" of mathematical operations.

13.09. Diverseness of Phonons

An additional reason to deny the phonon-photon analogy is the singlefacedness of photons against the diverseness and multifacedness of phonons. There are "thermal" phonons, which represent quanta of thermal vibrational energy. There are also "longitudinal" and "transverse" phonons, in acoustical and several "optical" branches.

In addition, there are rotons, helicons, and other phonon-like energy quanta in solids and liquids. Each of these "quasi-particles" has its own characteristic energy relations. Only acoustic phonons, in low and medium frequency ranges, have energies exactly proportional to frequencies, as derived in Section 7, and as postulated for photons by Planck's law. The behavior of other phonons has no exact parallel among electromagnetic waves in the vacuum space.

We can show that the real diverseness of phonons results from physical differences between the constituent particles in condensed matter, and from their closeness, not from mathematical presentations. In solids and liquids, there often is an overlap of the outer electron orbits of the constituents, even of electron shells. Therefore, while the interaction of constituents can be approximated in the "long-time average" as that of "material points", or "point charges", the atoms, ions, and molecules of solids and liquids "see" and "feel" each others' planetary structure, at least of the outermost shells.

The distributions of charge, mass, and magnetic moments in the particles depend at any moment on the direction, or "viewing angle". At the extremely small distances between the constituents, their distributions are not "spherically symmetric" to each other. They have their "faces" and "backs", hard "fists" and soft "bellies".

As it happens in dense crowds, the constituents of a solid "feel" the curvatures of each other, and respond to them, even to small rotations and "breathing" pulsations. The various kinds of phonons represent the energy quanta of each kind of displacement of the various <u>parts</u> of the constituent particles.

Contrarily, the distance between the bound electrons and positrons in space is 50 times larger than their radii. Therefore, independent of whether the yet unknown internal composition of electrons and positrons is that of closely packed matter, as follows from the identical density of all kinds of nuclear particles, or whether it is "diluted" as is the planetary structure of atomic matter,
> the bound electrons and positrons of space
> "see" each other and interact as "material points"
> or point charges *par excellence*, without any "side effects".

Thus, the photon represents a quantum of the "clean" wave energy of a bound electron or positron of space. The photon therefore depends only on the wave frequency, as postulated in Planck's law.

13.10. The Particle Features of Phonons and Photons

The particle features of photons are revealed in their interactions with electrons, e.g., in the photoelectric effect, in photoionization, etc. The most striking particle feature of the photon is its momentum. Momenta of photons were revealed in 1923 by A.H.Compton, in his experiments on the scattering of X-ray photons on 'free' electrons in graphite and in other materials.

The particle features of photons led to the consideration of light as streams of photons, i.e., to a certain revival of Newton's corpuscular theory of light. However, trials to use some consequences of Newton's theory, e.g., that the velocity of the corpuscles of light is larger in matter than in empty space, led to wrong results, and were abandoned. This happened to Einstein's assumption that the velocity of X-ray photons in matter is larger than in vacuum.

The particle features of PHONONS were revealed in the scattering of 'free' electrons in solids and liquids, e.g., in the scattering of conduction electrons in semiconductors and in metals. It turned out that the various scattering processes of electrons and holes, in collisions with the thermally and acoustically vibrating constituents of semiconductors and metals, are most successfully dealt with by treating these processes as particle collisions between electrons and phonons.

In these treatments, as well as in the related experiments, phonons also turned out to have momentum, i.e., the most striking particle feature. Moreover, the momenta of phonons were found to be significantly larger and easier to detect than the momenta of photons of visible light. The momenta of photons and phonons were also revealed in experiments, which could be treated as particle collisions between photons and phonons in solid and liquid bodies.

The particle features of the phonon result from its physical nature. The phonon represents the action of a single vibrating constituent particle of the body carrying the acoustic wave. When a small target, of the size of the constituent particle, is placed in the way of the wave, the target can be knocked by a single vibrating particle in the wave. The small target will therefore absorb the wave energy and momentum of this single particle. By this absorption, the target really detects a real vibrating particle, namely the constituent atom, ion, or molecule of the body which carries the wave.

Therefore, while using for the absorbed wave energy and momentum the name "phonon", one should not forget the real vibrating particle of the medium, which submitted the phonon to the target.

13.11. The "Particle-Wave Duality Principle" Deception

The particle features of photons are long considered as proof that electromagnetic waves have particle properties. This contradicts the fundamental distinction between the physical concept of a wave, and the physical concept of a particle. For example, waves can move or penetrate through one another, and diffract around obstacles, while particles cannot. Coherent waves propagating in the same direction can interfere destructively, up to cancelling one another, while beams of identical particles moving in the same direction can only be enhanced.

Another contradiction was added in 1927, when it was established that electron beams, scattered from crystal surfaces, undergo diffraction and destructive interference, similar to X-rays. The observed particle properties of electromagnetic waves and wave properties of electrons cannot be explained by any existing theory. Thus, as customary, quantum theory introduced a postulate, stating without any explanation, that
    waves may exhibit in some phenomena wave properties,
        and particle properties in other phenomena;
similarly,
    particles may exhibit in some phenomena
        regular particle properties,
        and wave properties in other phenomena.
This postulate is the "particle-wave duality principle".

It is easy to see that the duality principle, like other postulates of the quantum theory, does not explain a thing. It only outlines the observed facts in the form of a physical law, as if saying to the student or reader: "this is what it is, and as it should be".

The principle is quoted in the literature and in teaching as if it were **the physical reason** for the wave properties of particles, and for the particle properties of waves. Thus, generations of physicists were and are still brought up in such a belief. Try to ask, not only a good student, but also a teacher or a working physicist, why particles exhibit wave properties, or why waves exhibit particle properties. The answer will be: "BECAUSE of the particle-wave duality principle."

13.12. Nature Behind the Particle-Wave "Duality"

The bulk deformation wave in solids and liquids exhibits particle properties, because the wave is carried by vibrating atoms, ions, or molecules. Due to the binding between them, each constituent particle transfers to the next particle the wave energy and momentum, obtained from the previous particle.

When there is a small target on the path of this transfer process, i.e., on the path of the phonon, then the target is knocked by the last constituent particle in the row. Being knocked by a single particle in the wave, the target naturally reveals the particle properties *of this particle*. Hence, what we consider as the particle properties of the wave are actually the particle properties of constituent particles of the body, carrying the wave.

The wave properties of particles are actually the properties of the real deformation waves, which the moving particle creates around its path, by causing the constituent particles of the medium to vibrate. Lattice units, or volumes, entered by the moving particle, expand, and contract when the particle leaves.

As a result, the motion of a particle in a solid or liquid body is accompanied by two waves. <u>One</u> is the deformation wave, carried by the vibrating constituents of the solid or liquid medium. This wave propagates along the path of the moving particle with the velocity of sound, and can be described as a phonon stream. <u>Second</u> is the de Broglie wave, carried by the bound electrons and positrons of space. This wave propagates along the path of the moving particle with the velocity of light, and is describable as a photon stream.

Hence,
> real particles moving in a real ordered medium
>> create real accompanying waves in it;
>
> the detected properties of these accompanying waves
> are misinterpreted as belonging to the particle itself, i.e.,
>> as some mystical wave properties of the particle.

Also,
> real waves are carried by real vibrating particles;
> the detected properties of these particles,
> are misinterpreted as belonging to the waves, i.e., as some mystical particle properties of the waves.

We may say that
> in every real wave, and behind every phonon, or photon
> there are real vibrating particles;

and, *vice versa*,
> around every moving particle, there are real waves,
> with their phonons and photons.

This is the nature behind the particle-wave duality.

Chapter 14

VELOCITIES OF BULK DEFORMATION WAVES AND THE "MASS-ENERGY EQUIVALENCE" DECEPTION

14.01. Velocities of Molecules and Sound in Gases

The experimental values of the velocity of bulk waves and sound in gases depend inversely on the masses of their molecules, and increase with the pressure p and temperature T of the gas. In dry unpolluted air, at 0°C (T=273 K), and "normal atmospheric pressure", the measured value of the velocity v_s of bulk waves and sound is

$$v_s = 330 \text{ m/s} .$$

At these "normal conditions", the average distance between the centers of the molecules in any gas is ~33 angstrom. This is 10 times larger than in solids and liquids, and 10 to 15 times larger than the sizes of the molecules. Therefore, molecules in gases can travel freely and undisturbed, on hundreds of angstrom long "free paths", until they collide with other molecules.

Clean dry air, and many other gases, when not too strongly compressed, and kept well above the temperatures at which they can be liquefied, behave like the *"ideal gas"* of the *"kinetic theory of gases"*. On distance, the molecules of such gases do not attract or repel each other. On contact, they do not stick or "glue on" to each other. The molecules interact only during collisions, and their collisions are purely elastic. The colliding molecules conserve not only their overall momentum, but also the overall kinetic energy.

Calculated from the kinetic theory of gases, the velocity of sound in air should be larger by 18% than the calculated value of the mean thermal velocity of the molecules along one of the space axes. This velocity is 280 m/s, hence the calculated value of the velocity v_s of sound in air at normal conditions is 330 m/s, which is also the experimental value.

The mean thermal velocity v of the molecules of a gas, along one of

the three coordinate axes, as well as the velocity v_s of bulk waves and sound in gases, are both determined by THE SQUARE ROOTS OF THE QUOTIENTS, of either
  the gas pressure, divided by the density of the gas, or
  the thermal energy density, divided by the density of the gas, or
  the mean thermal energy of a molecule,
    divided by the mass of the molecule.
Each of these quotients is equal to the thermal energy per unit mass.

14.02. Propagation Mechanisms of Bulk Waves in Various Media

Bulk waves in gases consist of half-wave clusters, the volume of one containing more molecules than in equilibrium, while the other contains fewer molecules. This is similar to the numbers of excess and missing particles in the half-wave deformation clusters of bulk waves in solids and liquids.

However in solids and liquids, the excess particles are bound, as all other particles, by energies a hundred times larger than their mean thermal energies. Therefore, the excess particles remain within an angstrom-thin monolayer on the surface between the clusters. They just vibrate around their lattice sites, with an amplitude well below an angstrom, alternately shifting from one cluster to the other. In this vibrational motion, the particles do not leave their monolayer.

In gases, the excess molecules penetrate hundreds of angstroms deep into the clusters. Colliding there with other molecules, they move backwards or sideways, and do not propagate with the wave. It would take them several seconds to travel a few meters, while the wave crosses such distances in a hundredth of a second. This means that, e.g., the molecules of the air at the vocal chords of a speaking person do not reach the ear of the listener. The ears are reached by the half-waves of increased and reduced pressure, which do not carry the air molecules from where the sound was generated.

The propagation of bulk waves and sound in gases is started by the creation of differences in pressure, i.e., of differences in thermal energy densities between half-wave clusters. Under appropriate coherence and interference conditions between traveling and reflected rudimentary waves, the wave motion can be established. In it, the wave energy, i.e., the excess thermal energy of half-wave clusters, is transferred with the velocity of the wave (of sound), until fully absorbed, or "randomized", in the medium.

In solids and liquids, the propagation of bulk waves starts with the creation of differences or excesses in the number of particles and in the binding energy density between the half-wave clusters. Differences in thermal energy density between the clusters have very little effect, if any, on the establishment and propagation of the wave. They do affect, though, the absorption and "randomizing" of a "dying" wave. Once again, in an established wave, the wave energy, or the excess binding energy of half-wave clusters, is transferred with the velocity of the wave, or sound, until absorbed in the medium.

In gases, the molecules are not bound to each other, the binding energy is zero. The wave energy transferred by the molecules originates from the differences in thermal energy. Hence the velocity of waves and sound in gases is determined by the square root of the thermal energy per molecule, divided by the mass per molecule. In solids and liquids, the energy transferred in the wave originates from the differences in the total binding energy contained in the half-wave clusters, and it is easy to show that the velocity of sound in the bulks (far from surfaces and boundaries) is determined by the square root of the binding energy per constituent molecule, atom or ion, divided by the mass per constituent.

Assume that in the liquefied or solidified substance, the constituent particle (molecule) remains unchanged, i.e., that the mass per particle remains the same in the three aggregation states. However the per-particle binding energy density in the solid is a hundred times larger than the thermal energy density. Hence, the velocity of bulk waves and sound in solids should be around ten (square root of 100) times larger than in gases. In liquids, the factor is ~5. This approximation agrees with the measured values of sound velocities, which are usually around 300 m/s in gases, 3000 m/s in the bulks of solids, and around 1500 m/s in liquids.

14.03. Velocity of Bulk Waves and Sound in Rocksalt Crystals

Let us consider a sodium chloride (rocksalt, NaCl) crystal so large, that the outer surfaces or boundaries of the crystal do not substantially affect the propagation of bulk waves in it. For this process, the crystal may thus be considered "unbounded". The crystal is also so large, that as large as its single-crystal grains may be, the conditions for the propagation of sound in it are identical in all directions. For this process, the crystal is "polycrystalline", and represents a uniform elastic medium.

In the lattices of sodium chloride and other alkali halides, there is one ion per unit cube. Hence, the binding energy density, or energy per unit volume, is the average per-ion (or per unit cube) binding energy $_bE$, divided by the volume l_o^3 of the unit cube. The (mass-) density is the mass m_i per average ion (or per unit cube), divided by the volume of the unit cube.

The quotient of the binding energy density by the (mass-) density is, therefore,

$$(_bE/l_o^3)/(m_i/l_o^3) = {_bE}/m_i .$$

Thus, in face-centered cubic lattices, the quotient is equal to the binding energy of an average ion, divided by its mass. The velocity v_s of bulk waves and sound in the face-centered cubic alkali halide crystals should be equal to the square root of this ratio, thus,

$$v_s = (_bE/m_i)^{1/2} .$$

In sodium chloride, the binding energy of an ion pair is 8 eV, hence $_bE = 4$ eV. The mass of the sodium (Na^+) ion is 23 amu, and of the chlorine (Cl^-) ion is 35 amu, thus

$$m_i = (23 \text{ amu} + 35 \text{ amu})/2 = 29 \text{ amu}.$$

In SI units,

$$m_i = 29 \cdot 1.66 \cdot 10^{-27} \text{ kg} = 4.8 \cdot 10^{-26} \text{ kg} ,$$

and

$$_bE = 4 \text{ eV} = 4 \cdot 1.6 \cdot 10^{-19} \text{ J} = 6.4 \cdot 10^{-19} \text{ joule}.$$

Substituting these data to the v_s formula, we obtain

$$v_s = (6.4 \cdot 10^{-19} \text{ joule}/4.8 \cdot 10^{-26} \text{ kg})^{1/2} =$$
$$= 3600 \text{ m/s} = 3.6 \text{ km/s} .$$

This value of 3.6 km/s is exactly the experimental value of the sound velocity in polycrystalline sodium chloride.

14.04. Velocities of Sound in NaCl Single Crystals

In large NaCl single crystals, in which the effects of the outer surfaces or boundaries can be disregarded, there are still six different values of the sound velocity, depending on the
- direction of propagation, relative to the edge,
- or face, or cube diagonals of the lattice,

and on the
- main direction of ion vibrations:
    - longitudinal or transverse,
- relative to the propagation direction of the wave.

These six experimental values are presented in Table 5.

TABLE 5

Measured Sound Velocities in NaCl Single Crystals

Direction of Propagation	Velocity in km/s of	
	longitudinal	transverse
	sound	
Along cube edges	4.74	2.41
along face diagonals	4.72	2.90
along cube diagonals	4.37	2.45

In a very large polycrystalline NaCl block, the ions vibrate in each half-wave cluster in all possible directions, and the single crystal grains are oriented randomly. Therefore, all directional differences are averaged out, and there should be no preferred direction of either particle vibrations, or wave propagation. The velocity of sound in this uniform elastic medium should thus be equal to the average of all six values listed in the Table. This average is

$$(4.74+4.72+4.37+2.41+2.90+2.45)(km/s)/6 = (21.59 \text{ km/s})/6 =$$
$$= 3.60 \text{ km/s} .$$

The obtained 3.6 km/s average value is exactly the experimental value of the velocity of sound in an unbounded polycrystalline NaCl sample, and exactly the value calculated with our formula for v_s. This proves the validity of our assumptions, not only for the NaCl crystal, but for any ionic fcc solid. Our assumptions may thus be applied to other bodies, in a more or less general way.

14.05. Velocity of Bulk Deformation Waves in the Epola and the Velocity of Light

Considering the electron positron lattice (epola) in space as being face-centered cubic (fcc), we may calculate the velocity v_d of bulk deformation waves in the epola, using the general formula

$$v_d = (_bE/m_e)^{1/2}.$$

Substituting the mass m_e of the electron or of the positron,

$$m_e = 9.1 \cdot 10^{-31} \text{ kg},$$

and the binding energy per particle, which is

$$_bE = 511,000 \text{ eV}, \quad \text{or } 8.2 \cdot 10^{-14} \text{ joule},$$

we obtain,

$$v_d = (8.2 \cdot 10^{-14} \text{ J} / 9.1 \cdot 10^{-31} \text{ kg})^{1/2} = (9 \cdot 10^{16} \text{ m}^2/\text{s}^2)^{1/2} =$$
$$= 3 \cdot 10^8 \text{ m/s} = 300,000 \text{ km/s} = \underline{c}.$$

Hence, the velocity of bulk deformation waves in the epola is identical to the velocity of light "in free space"., i.e., to the velocity of electromagnetic radiation. Our conclusion is, therefore, that
 what we detect as light, and as electromagnetic radiation,
  starting from the longest radio waves of lowest frequencies,
  and ending with the most energetic gamma rays,
   of shortest possible wavelength, ($w = 2l_o = 8.8$ fm),
is due to the propagation of bulk deformation waves in the epola.

The identity of velocities does not mean that electromagnetic waves are identical to the bulk deformation waves of the same frequency. One of the most striking distinctions is that electromagnetic waves are detected as TRANSVERSE waves, while in the bulk deformation waves the constituent particles of the medium vibrate in all possible directions, with no preference to any particular direction. This, and other distinctions, are also explained by the epola structure of space.

14.06. Energy of Freeing and Capture of Electron Positron Pairs

The formula of the velocity of bulk deformation waves in the epola,

$$v_d = c = (_bE/m_e)^{1/2},$$

can be rearranged for the calculation of the binding energy $_bE$ of an electron or positron in the epola, to obtain,

$$_bE = m_e \cdot c^2.$$

Hence, the binding energy per particle in the epola is equal to the mass of the particle, multiplied by the squared velocity of light.

This formula can be used for the calculation of the energy, needed to free electron-positron pairs out of their bonds in the lattice. The freeing of a single particle would violate the electric neutrality of the lattice, and is possible under special conditions, like a "knockout" by a very fast nuclear particle.

Multiplying both sides of the formula by an even number n, we obtain,

$$n \cdot _bE = n \cdot m_e c^2.$$

This formula yields the binding energy, or the freeing energy of an even number n of electrons and positrons.

Let us now replace $n \cdot _bE$ by E, (meaning that $E = n \cdot _bE$), and $n \cdot m_e$ by m (meaning that $m = n \cdot m_e$). We obtain,

$$E = mc^2,$$

which reads:
    the energy E, absorbed to free a mass m of bound particles
        <u>out</u> of their bonds in the epola,
or
    the energy E, emitted when a mass m of free particles
    is captured <u>into</u> the bonds of the epola
    is equal to the mass m of the freed or captured particles,
multiplied by the squared velocity of light.

Note that our derivation of the $E=mc^2$ formula is achieved within a few lines of very simple algebra, as opposed to the heavy mathematical canons, used by Einstein (Lorentz transforms, heavy calculus, etc.). It is always so, that based on the right physical model, one can derive the needed results with much less mathematics.

Note also, that in our presentation, the $E=mc^2$ formula does not express a "mass-energy equivalence", just as the Anderson experiments do not represent any creation or annihilation of electrons and positrons, or any turning of mass into energy, or of energy into mass. The formula is just a mass-energy relation for the capture or freeing of particles, into or out of bonds in the epola.

14.07. Mass-Energy "Equivalence" in Ionic fcc Crystals

Let us think of a researcher who does not know the "exact" structure of the sodium chloride crystal (many don't!), or does not want to know (too many don't! And there is no time or money for knowing).

He discovers that under illumination by ultraviolet radiation, the transparent and insulating crystal becomes not so transparent, and somehow conducting. Analyzing the nature of the conductivity, he finds that it is due to the appearance of mobile or 'free' Na^+ and Cl^- ion pairs, one per every absorbed 8 electron·volt photon. The presence of these ions he could not detect, by any of his means, neither before, nor late after the illumination. Hence, the researcher concludes that the ions were "created" by the illumination.

The researcher finds then, that at the ceasing of illumination, and during a certain time thereafter, the crystal emits light, also in the ultraviolet, though never in the 8 eV range. Thereafter, the crystal returns to its transparent and insulating state, with no more ions in it. Using many assumptions, and exploiting numerous correction factors, the researcher concludes that the ion pairs "created" under illumination, gradually disappear or "annihilate" afterwards, and that with the disappearance of each pair, the crystal emits 8 eV of energy, but in at least two photons, or a photon and phonon(s).

Having made these perfect and exact experimental discoveries, which are quite analogous to C.D.Anderson's 1932 discovery of the electron-positron "creation" and "annihilation", our researcher is looking for a theory "to fit". And what is then better, more publishable and promissory of funding, than experimental results, attributed to

Einstein's mass-energy equivalence?

Thus, our researcher starts to substitute his results to Einstein's $E=mc^2$ formula, and writes

$$8 \text{ eV} = 58 \text{ amu } (3 \cdot 10^8 \text{ m/s})^2 = 58 \cdot 1.66 \cdot 10^{-27} \text{ kg} \cdot 9 \cdot 10^{16} \text{ m}^2/\text{s}^2 =$$
$$= 8.7 \cdot 10^{-9} \text{ J} = 8.7 \cdot 10^{-9} \cdot 6.25 \cdot 10^{18} \text{ eV} =$$
$$= 5.4 \cdot 10^{10} \text{ eV} .$$

Alas! The right-hand side of the substitution yields 54 billion eV, which is not the expected 8 eV, "standing" in the left hand side of the equation. Hence, the result obtained contradicts the teachings of the Master, that *"The mass of a body is a measure of its energy content; if the energy changes by E, the mass changes in the same sense by E/c^2."*

Thinking of what is wrong, our researcher asks himself why he has to use the squared velocity of light in free space, when the experiments were performed in rocksalt. He tries to use the squared velocity of light in rocksalt but still gets 100 million times larger energies than 8 eV. Finally, he tries the velocity of sound, not of light, and substitutes:

$$8 \text{ eV} = 58 \text{ amu} \cdot (3600 \text{ m/s})^2 = 58 \cdot 1.66 \cdot 10^{-27} \text{ kg} \cdot 1.3 \cdot 10^7 \text{ m}^2/\text{s}^2 =$$
$$= 1.25 \cdot 10^{-18} \text{ J} = 1.25 \cdot 10^{-18} \cdot 6.25 \cdot 10^{18} \text{ eV} =$$
$$= 8 \text{ eV} !$$

Eureka, he got it! The right-hand side is 8 eV, equal to the left side. Now our researcher generalizes his achievement by writing the "fitting" formula

$$E = m \, v_s^2,$$

which he considers as *"the mass-energy equivalence law"* for the NaCl and similar crystals. Based on his results, and on his scholarly upbringing, he firmly believes that the absorption of 8 eV in the crystal really creates in it the two ions, of mass 58 amu, and that their annihilation releases the equivalent energy of 8 eV. He then goes farther, (as did Einstein in 1911, and as physicists still do), and states that energy has mass, and that the mass of 8 eV is 58 amu.

14.08. The Mass-Energy Equivalence Deception

To us unbelievers, it is clear that 8 eV cannot create any mass, let alone 58 amu. It is also clear, since the times of A.L.Lavoisier, that even the smallest observable mass cannot be annihilated, or turned into energy. (No, Lavoisier was beheaded not for establishing the anti-Einsteinian Law of Mass Conservation.)

Bound masses cannot be detected by regular means, but appear to these means when freed out of bonds. This freeing should not be interpreted as a creation of masses, though it may be legitimately calculated in such a way. The same applies also to the freeing and capturing of electron-positron pairs in space, with absorption and emission of 1.02 MeV.

The $E=mc^2$ formula is considered as the greatest achievement of modern science, especially by populations, anyhow unable to tell the difference between mass and energy. We obtained both the $E=mc^2$ equation and the $E=mv_s^2$ equation by rearranging the formula for the velocity of bulk deformation waves in the epola or in ionic face-centered cubic lattices, and by substituting in them the mass and binding energy of the lattice particles.

To us, the $E=mc^2$ and $E=mv_s^2$ equations have exactly as much to do with real mass-energy equivalence, as does the $E=mv^2/2$ formula for the kinetic energy (i.e., nothing). We conclude that in both the epola "free space", and in the ionic fcc lattices,
  the mass-energy "equivalence" formulas
  result from the structure of the media;
  they express energy relations
  for the freeing of masses out of bonds
  and for their capture into the bonds,
but not for any real creation or annihilation of a mass.

Chapter 15
SPECTRAL COMPOSITION AND NATURE OF LIGHT

15.01. The Spectral Composition of Light

Newton was the first to prove that sunlight, and "white" light in general, is composed of rays having all the colors of a rainbow ("Optiks", 1672). In his arrangement, sunlight passing through a horizontal slit in a window shade enters a dark room, and is incident on a "base up" glass prism. After two refractions, into the prism and out of it, the light emerges, bent upwards (toward the base of the prism), and falls on the opposite wall, which represents a white screen. On the screen was a continuous succession of overlapping images of the slit, in all colors of the rainbow. The uppermost image was violet, followed downwards by blue, azure, green, yellow, orange, and red images, with no sharp borders between them.

The pattern obtained by separation of radiation into its components, is the "spectrum" of the radiation. In the spectrum described, the separation was due to the different strengths with which the prism bends these rays, or to the different "indices of refraction" in glass for lights of different colors. Violet light is bent strongest, its refractive index in glass is largest; red light has the smallest refractive index in glass, and is bent least.

The separation of radiations due to their differing indices of refraction in a material is named "dispersion". Radiation spectra obtained due to dispersion are "dispersion spectra". In them, the different radiations are arranged in order of their refractive indices. Of known materials, diamonds have the largest refractive indices, and the largest dispersion, i.e., the largest differences in refractive indices for lights of different colors. This makes them so beautifully sparkling, when properly cut, and rich in colors, to be "girls' best friends".

It is amazing to note that it took science 128 years before the possibility of the presence of radiation beyond the visible spectrum was considered. In 1800, F.W.Herschel placed a thermometer below the red light of the spectrum, and found heating there. The invisible radiation which caused the heating was named "infra-red", which means "below red". A year later, J.W.Ritter and, independently, W.H.Wollaston found (employing a quartz prism) that above the violet range, there is a non-heating but chemically active invisible radiation. This radiation was named "ultra-violet" meaning "above violet". (Quartz is transparent to the ultraviolet radiation, while glass isn't).

It then took decades to prove that the infrared and ultraviolet radiations propagate with the velocity of light, undergo dispersion, diffraction, and interference, as does visible light, so that they all represent branches or ranges of one kind of radiation.

15.02. Continuous and Line Spectra of Light

The spectrum of sunlight, observed by Newton, appeared continuous, i.e., without border lines between the different color regions, and without visible dark or bright lines. However in 1814, J.Fraunhofer, using his precise dispersion spectroscope, obtained a widely expanded and bright sunlight spectrum, with numerous sharp dark lines in it.

These "Fraunhofer lines" show that the radiation which produced the lines, was absorbed on its way from the hot surface of the Sun by the cooler gases in the Sun's atmosphere. The lines were later identified as the characteristic lines, constituting the spectra of hot gases, such as hydrogen, helium, and of other elements.

It turned out that the spectra of light, emitted by equally hot solids, liquids, and highly compressed gases are continuous, and undistinguishable from one another. Continuous spectra do not depend on the chemical composition of their sources. But the luminosity (brightness) of the spectra depends on the geometry and structure of the emitting surface, and on the temperature of the source.

Contrarily, the spectra of light, emitted by uncompressed hot gases are "line spectra". In the spectroscope, they show up as a black or dark background, with several sharp shiny lines on it. These lines are characteristic of each particular gas; they are its "fingerprints" or "signatures". They allow one to perform a most precise "spectral analysis" of the chemical composition of the radiation source.

The number of lines in the spectra of elementary monatomic gases is usually the larger the higher the number of orbital electrons in the atom. In molecular gases, the number of lines depends also on the molecular structure. When the molecules of the gas or vapor consist of many atoms, or when the gas or vapor is slightly compressed, then the lines (or some lines) of the spectrum "coagulate" into wider "bands", and the line spectrum may turn into a "band spectrum". Again, the higher the temperature of the source the higher is the luminosity of its spectrum.

Fraunhofer lines, observed on the background of a continuous spectrum, indicate that the source of the radiation must be a hot solid, liquid, or highly compressed gas. However on its way to the spectroscope, the radiation of this source must be passing a layer (or layers) of uncompressed gases. If the lines are dark, then this means that the gases absorb, within their "signature" lines, more radiation energy from the source than they emit. Hence, these gases are cooler than the source. Such spectra are thus "absorption spectra".

If the lines are brighter than the background, then the temperature of the gases is higher than the temperature of the source, whose light produced the continuous spectrum. These "emission spectra", as well as the absorption spectra, can be used for chemical analyses of the gases in-between (e.g., of the gases in the atmosphere of the Sun).

15.03. Interference of Light Waves, and Interference Spectrometry

J.Fraunhofer made another breakthrough, when he developed (in 1821) the "diffraction grating" and replaced with it the prism in the spectroscope. He thus created the "interference spectrometer", based on the phenomena of light diffraction and interference.

Diffraction and interference are wave phenomena *par excellence*. When a wave reaches an obstacle, on which it is stopped, then part of the wave energy bends or diffracts around the edges of the obstacle. As a result of diffraction, the wave motion continues also behind the non-transparent obstacle.

A spot at a certain distance behind the obstacle can be reached by waves which diffracted around opposite edges of the obstacle. The diffracted waves then interfere with each other in every spot reached by them. It is clear that the distances, crossed by the diffracted waves, from "their edge" of the obstacle, to a given spot, are not necessarily equal to one another.

It may thus occur that a certain spot is reached at a certain moment by one diffracted wave at a phase, commanding the particles in the spot to shift upwards, while the other wave commands an amplitudinal displacement downward. As a result, the deflection of the spot is reduced, and can even be zero. In another spot, the diffracted waves may arrive at the same or similar phase, issuing the same or similar commands to the particles of the spot. As a result, the deflection of the particles in this spot is increased.

The rule by which one can find the result of the interference of diffracted (or other COHERENT) waves is, that
    if the difference in the paths of the waves,
        from the edges of the obstacle
        to the spot under consideration,
    contains an even number of half-waves,
        then the waves reach the spot at the same phase;
    their interference in this spot is CONSTRUCTIVE,
and increases the amplitude and wave energy in this spot.

If the path lengths of the diffracted waves, from the edges of the obstacle, to a given spot,
    differ by an odd number of half-waves, then
    in this spot, the two waves meet at opposite phases
(e.g., one wave commands "up", while the other commands "down");
    the interference in this spot is DESTRUCTIVE;
it reduces the amplitude and the wave energy, even to zero.

The diffraction grating consists of a large number of obstacles to light that are much thinner than hair, placed at small and equal distances from one another, e.g., ten thousand per inch. Hence, an inch-wide bundle of light-waves, crossing the diffraction grating, produces an interference picture that is by ten thousand times stronger than one obstacle. Therefore, "interference spectrometry" made it possible to determine for each spectral line the corresponding wavelength and frequency of radiation, with an accuracy of one part per ten million!

15.04. Wavelengths and Frequencies of Visible Light

The wavelength of the "lowermost" red light was measured to be 0.8 microns (micrometers, μm), or 800 nanometers (nm), and of the "uppermost" violet light, 0.4 μm, or 400 nm. To find the frequency f of the waves, we should divide the velocity c of the wave by its wavelength w,

$$f = c/w .$$

Hence, for the "lowermost" red light we have,

$$f = (3 \cdot 10^8 \text{ m/s})/8 \cdot 10^{-7} \text{ m} = 3.7 \cdot 10^{14} \text{ s}^{-1} =$$
$$= 3.7 \cdot 10^{14} \text{ Hz (hertz)},$$

or 370 THz (terahertz). The "uppermost" violet light has a wavelength that is two times shorter, thus a frequency of 750 THz, that is two times higher. Therefore, the wavelengths of visible radiation vary
from 400 nm (violet) to 800 nm (red),
and the frequencies vary between 370 and 750 THz (terahertz).

The spectrum of hydrogen in the visible range contains four prominent lines,

      an orange line,   of w = 656 nm, and f = 457 THz,
      an azure line,   of w = 486 nm, and f = 617 THz,
      a blue line,   of w = 434 nm, and f = 691 THz,
and   a violet line,   of w = 410 nm, and f = 731 THz.

The spectrum of helium has 7 prominent lines in the visible region.

15.05. Early Theories of Light

The development of interference spectrometry marked the beginning of the end of Newton's "corpuscular" theory of light, which ruled in science for 140 years. The main reason was that corpuscles, meant to be very small bodies of atomic matter, cannot undergo interference, the way waves do. They also cannot move through one another, as waves do.

When two beams of atomic bodies meet at a small angle, then they can only enhance each other. Hence, the result of the interference of corpuscular beams moving in similar directions in a Newtonian absolutely empty space can be "enhancement only". Contrarily, two similar waves enhance each other only in spots where they meet at equal phases. In spots, where they meet having opposite phases, the two waves may cancel each other. Hence,

if the interference of two similar beams
results in their cancellation at certain spots in space,
then these beams must consist of waves,
not of corpuscles of atomic bodies.

Newton's corpuscular theory of light, developed in the book "Optiks" (1672), was connected with his belief in the absolute emptiness of space *per se*. Unlike waves, which have to be carried by some elastic ambient ("ether"), corpuscular beams do not need an ambient for their propagation. Newton's corpuscular theory led him to an explanation of the refraction of light in transparent bodies as due to the attraction of the corpuscles by the particles of these bodies. Hence, according to the corpuscular theory, the velocity of light in materials is increased; it is larger in them than in the empty space by a factor, equal to the refractive index of the material.

For example, if the refractive index of water is 1.33, and of glass 1.55, then Newton and his followers could prove, mathematically of course, that the velocity of light in water is 1.33 times larger than in the empty space (or air, approximately), and in glass it is 1.55 times higher than in space. During the 140 years of the absolute reign of Newton's theory, there were indications that this is wrong and that the velocity of light in materials is smaller than in space. However in science, as in politics, it is not easy, and often impossible to turn down the false, if it was presented by an accepted authority.

The wave theory of light was introduced in 1678 by C.Huygens. He used to send some works to the Royal Society in London, but figured that after Newton's "Optiks", his contradicting theory will have no chance there. He thus presented the book (*"Traité de la lumiere"*) to the French Academy. However, in spite of their animosity to the British, they still delayed the publication for 12 years, until 1690.

With his wave theory, Huygens could account for all the facts about light known then. In particular, he derived the laws of reflection and refraction of light, with the refractive index of materials showing how many times the velocity of light in them is <u>smaller</u> than in space. Strangely, Huygens did not mention dispersion, nor did he relate to Newton's separation of sunlight into its composite rays, though these phenomena can best be explained by the wave theory.

As was experimentally proven during the XIX century, the velocity of light in atomic bodies is always smaller than in space, and the refractive index shows how many times smaller it is. The discovery of the diffraction, interference, and of the polarization of light, have proven by 1835, already beyond any doubt, that light represents wave processes in space and in atomic bodies.

15.06. The Electromagnetic Theory of Light

A great step forward was made by J.C.Maxwell, who created in 1865 the electromagnetic theory of light. Maxwell's Theory is based on the enormous progress in electromagnetism, achieved by Michael Faraday, who denied Newton's empty space and "action on distance". Maxwell concluded that any accelerated motion of electrical charges, motions of magnetic bodies, and changes in their magnetic moments, must excite electromagnetic waves in the surrounding "ether substance", considered as the carrier of the electromagnetic field.

By Maxwell, all electromagnetic waves propagate in the ether with the velocity of light c, and differ from one another by their frequencies. The frequencies of the electromagnetic waves depend on the velocities and frequencies of the electric and magnetic changes or processes, due to which the waves were generated. Thus, light represents electromagnetic waves, generated by intra-atomic and intramolecular motions of electric charges and magnetic moments.

A crucial experimental proof of Maxwell's electromagnetic theory was provided with the discovery of radio waves by H.Hertz in 1887. These "Hertzian waves" were generated by electric spark discharges. Their frequencies were in the range of millions per second, or MHz (mega-hertz). The frequencies of the waves could be changed, e.g., by regulating the spark gap, but the velocity of their propagation in space (and in air) was always the velocity of light.

The electromagnetic theory suffered a serious blow when the results of the Michelson-Morley experiments of 1887, as well as the improved versions of 1904, up to 1930, denied the existence of the hypothetical massless "luminiferous ether". This marked a return to

Newton's empty space, especially after the postulation of the emptiness of space by A.Einstein in 1905 and his arguing that the velocity of X-rays in materials is higher than in the empty space.

There were also several indications of the discrete nature of radiation, which no wave theory could fully account for. The most striking was the very existence of discrete line spectra. Then came the 1887 discovery of the photoelectric effect. In hard metals, the effect occurred only under ultraviolet illumination. The effect could not be obtained in them even with the brightest and most powerful beams of visible light. No wave theory can explain such a behavior.

The electromagnetic theory of light, together with a branch of physics called thermodynamics, yielded important laws for the electromagnetic radiation of a "blackbody", i.e., of a radiating cavity in a hot solid body. However these laws were unable to fit the short wavelength range in the spectra of the "blackbody radiation".

15.07. Planck's Quantum Postulate

In 1900, Max Planck derived an equation which fitted both the long- and short wavelength ranges of the blackbody radiation. The equation became the successful "Planck's Spectral Distribution Law". To derive the law, Planck had to introduce a postulate, which contradicted the then established physical concepts. Planck's postulate stated that

the radiation energy of a blackbody
is emitted in discrete amounts or "quanta" of energy;

the energy E_p of a radiation quantum, or photon
is proportional to the frequency f of the radiation.

Thus, the energy of a radiation quantum, or photon, is

$$E_p = h f .$$

Here h is a proportionality coefficient, showing the hypothetical "photon energy" of radiation, the frequency of which would be 1 Hz. The value of h was laboriously derived by Planck, and is "Planck's constant"; in SI units,

$$h = 6.63 \; 10^{-34} \text{ J/Hz , (or joule·seconds).}$$

If energies are expressed in electron·volts (eV), then
$$h = 4.14 \cdot 10^{-15} \text{ eV/Hz},$$
or 4.14 femtoelectron·volts/hertz, also 4.14 feV·s.

For example, the photon of red light, of frequency $f = 3.7 \cdot 10^{14}$ Hz, has energy
$$E_p = (4.14 \cdot 10^{-15} \text{ eV/Hz}) \cdot 3.7 \cdot 10^{14} \text{ Hz} = 1.5 \text{ eV}.$$
The photon of violet light, of frequency $f = 7.5 \cdot 10^{14}$ Hz, has energy
$$E_p = (4.14 \cdot 10^{-15} \text{ eV/Hz}) \, 7.5 \cdot 10^{14} \text{ Hz} = 3.1 \text{ eV}.$$
Thus, photon energies of visible light range between
$$1.5 \text{ eV (red) and } 3.1 \text{ eV (violet)}.$$

Planck's postulate does not explain what in the nature of radiation could cause the energy to be quantized. Nor does the postulate explain, what in the nature of radiation causes the quantum of its energy to be proportional to the frequency, while the vibrational energy of a vibrating body (oscillator) or spot in a wave is proportional to the <u>square of the amplitude</u>.

Because of these two unexplained contradictory statements, Planck had to ask his colleagues to let him through with his postulate, without necessarily accepting it, and then see where it leads to. They did, mostly because they considered the postulate a gimmick, applicable only to the derivation of the spectral distribution law. It turned out that the never explained postulate itself became an unexplained cornerstone of science, known as Planck's Law.

Letting Planck through with his postulate would not occur nowadays. If in our enlightened times a statement contradicts "established scientific knowledge", especially the established mathematical derivations, then referees and publishers reject the presentation outright, without bothering to find out if it leads to anything of value.

15.08. Einstein's Theory of the Photoelectric Effect

Planck's quantum postulate was used by A.Einstein in 1905 to explain the photoelectric effect. This effect was discovered in 1887 by H.Hertz, who observed that shining ultraviolet light on metal electrodes caused their electrical charges to decrease. Twelve years later, P.Lenard found that the decrease is due to the emission of free electrons out of the illuminated surfaces of metals. Lenard also established the experimental rules of the photoelectric effect. However these rules could not be explained by a wave theory of light.

Einstein postulated that light is not only emitted in discrete quanta or photons, but also interacts with electrons as a stream of discrete photon particles. In the photoelectric effect, the photons interact individually, each with one electron, which is 'free' inside the metal. The energy of the photon is fully (or not at all) transferred to the electron. To get out, the 'free' electron must overcome an energy barrier on the surface of the metal, losing on it an "exit energy", named (unfortunately) the "work function" of the metal.

If the energy of the 'free' electron is smaller than the work function, then the electron cannot leave the metal. Hence the work function can be measured as the minimum or "threshold" energy of photons, able to cause the photoelectric effect. The energy E of the "photoelectron", i.e., of the electron, which was successfully freed out of the metal, is smaller than the photon energy $E_p=hf$ by the value of the work function W of the metal,

$$E = E_p - W = hf - W.$$

This is Einstein's equation for the photoelectric effect.

The work functions of many metals: iron, copper, silver, tungsten, are around 4.5 eV; largest is the 5.4 eV work function of platinum. The smallest work functions are those of alkali metals: 1.9 eV in cesium, 2.3 eV in sodium. The photoelectric effect in alkali metals (and in barium) can thus be caused by visible light.

15.09. Questions on the Nature of Photons and Light

Einstein's Nobel prize winning theory of the photoelectric effect does not help one to understand the nature of photons. The same questions which were asked regarding Planck's postulate, as to why radiation is quantized, and why the photon energy is proportional to the frequency, remain unanswered, and new unanswered questions arise. If light is a flux of photon particles, then how can the definite wave features of light be explained? And if photons are particles, then of what? Are they particles of "empty space" proclaimed by Einstein to be the carrier of radiation? But then, the knocking of an electron out of a metal by a particle of emptiness sounds like a new problem.

The ability of photons to knock out electrons is also revealed in a number of phenomena, discovered after the photoelectric effect. Ultraviolet photons in the 10 eV range can knock atomic valence electrons out of their atoms, turning the atoms into ions. The "ionization" of atoms by photons is named "photoionization". For example, the ionization energy of a hydrogen atom is 13.6 eV.

Photons of the more energetic "far ultraviolet" can knock out electrons from deeper energy levels in atoms. X- rays, discovered in 1895, have photons of energies from 100 eV to 100,000 eV, and can knock out electrons from the deepest levels in the heavy atoms. The most energetic are gamma rays, first discovered in radioactive decays, and later in cosmic radiation; their photons have energies from 0.1 MeV to over 100 MeV. They can knock electrons and positrons out of the "empty space" itself (C.D.Anderson, 1932), or initiate and cause nuclear reactions. Can such powerful particles then be "particles of emptiness"?

Amazingly, A.Einstein in 1911, also in 1916 and later, considering the propagation of X rays in materials, with a refractive index, seeming to be smaller than unity, came out with the Newton-like argumentation that the propagation velocity of X-rays in materials is larger than their velocity in space. This argumentation survived until the 1950s; it was presented as a revival of Newton's corpuscular theory which, though inapplicable to light, is right for X rays. Only later it turned out that it is wrong for X-rays, too, and this kind of corpuscular handling was abandoned.

All these radiation phenomena and problems cannot be understood without knowing the physical nature of photons. They are all explained and answered by taking into consideration the electron-positron structure of space.

Chapter 16

STABILITY AND RADIATION OF ATOMS

16.01. Balance of Forces and Energies in the Hydrogen Atom

The atom of hydrogen is the simplest of all atoms. Its nucleus is just a single proton, the positive +e charge of which is neutralized by the negative charge -e of the single orbital electron of this atom. The orbital electron is moving on its stable orbit, due to the electrostatic and the inertial interactions with the surrounding epola space.

The motion of the orbital electron is analogous to the orbital motion of a planet. The planet, too, moves on its stable orbit due to two interactions: the gravitational and the inertial interactions with the surrounding space. Because of this analogy, the structure of atoms, discovered by E. Rutherford in 1911, is considered "planetary". The analogy is limited to the balance of forces and energies of the orbiting electrons and planets.

Because of the two interactions, there are two forces acting on the orbiting body: the attractive force (electrostatic or gravitational) toward the central body of the system (atomic nucleus, the Sun), and the inertial force. Without the attractive force, the orbiting body would be moving by inertia away from the central body. The planet, or electron, would then be moving on a straight path, tangent to the orbit, with a constant velocity v, which the body had at the moment when the attractive force was "switched off". Without inertia, on the other hand, the planet, or electron, would be moving toward the Sun, or the atomic nucleus, i.e., "falling" on it, and collapsing with the central body.

The energy needed to tear off the electron from the hydrogen atom in its normal, non-excited, or "ground state", is 13.6 eV. This is the **ionization energy** of the hydrogen atom. This energy must be equal to the energy which is binding the electron to the nucleus (or to the rest of the atom). The binding energy is the per particle energy of this system, i.e., half of the electrostatic energy of the system.

On the other hand, the binding energy must be equal to the energy of the inertial motion of the electron on the orbit, which is the kinetic energy. When the kinetic energy of the electron on an orbit is larger than the binding energy, then the electron moves away to a more remote orbit. When the kinetic energy is smaller than the binding energy on a given orbit, the electron "falls" to an orbit closer to the nucleus.

Hence, the kinetic energy of the electron on a _stable_ orbit must be equal to the binding energy, i.e., to half the electrostatic energy of the system. If not, then the electron cannot remain on the orbit. This "energy balance" is expressed by the equation,

$$m_e v^2/2 = ke^2/2R .$$

The left-hand side of this equation is the kinetic energy of the electron, and the right-hand side is half the electrostatic energy of the system, which is the binding energy of the electron in the atom.

If in any point of the orbit, the kinetic energy of the electron becomes larger than the binding energy, e.g., due to the action of an external force, then the electron moves by inertia away from the nucleus. In this motion, usually on a spiral, the electron performs work against the attractive electrostatic force, on account of its kinetic energy. Hence, the kinetic energy of the runaway electron gradually decreases with increasing distance R from the nucleus.

The runaway motion continues until a certain distance R from the nucleus, where the decreasing kinetic energy becomes equal to the binding energy there. The binding energy on an orbit, i.e., half of the electrostatic attractive energy, also decreases with increasing radius, or with the distance from the nucleus (Coulomb's Law).

If the kinetic energy of an orbital electron becomes smaller than the binding energy, e.g., due to an external force, then the electron is accelerated toward the nucleus, usually on a spiral. The velocity and kinetic energy of the electron are growing, and so also is the attractive electrostatic energy. The motion continues, until at a certain distance R from the nucleus, the increased kinetic energy becomes equal to the increased binding energy (1/2 the increased value of the electrostatic energy), and a new orbit is formed.

It is important to remember that the larger the radius R of an orbit, the smaller the attractive electrostatic energy of the orbital electron, and the smaller its binding energy, velocity, and kinetic energy. *Vice versa*, the smaller the radius of the stable orbit, the larger the attractive electrostatic energy of the orbital electron, and the larger its binding energy, velocity and kinetic energy.

16.02. Calculation of the Orbital Velocity in the Hydrogen Atom

The known value of the ionization energy of the hydrogen atom is 13.6 eV (electron·volt), or, in SI units (joules),

$$13.6 \cdot 1.6 \cdot 10^{-19} \text{ joule} = 2.17 \cdot 10^{-18} \text{ J}.$$

The kinetic energy $_kE$ of the electron on the orbit is

$$_kE = m_e v^2/2.$$

Solving this formula for the calculation of the orbital velocity v, yields,

$$v = (2 \cdot {_kE}/m_e)^{1/2}.$$

It follows from the energy balance, that the orbital kinetic energy of the electron is equal to the binding energy. On the "normal" or ground state orbit, the binding energy is equal to the ionization energy. Substituting for $_kE$ the equal value of the ionization energy, we obtain

$$v = (2 \cdot 2.17 \cdot 10^{-18} \text{ J}/ 9.1 \cdot 10^{-31} \text{ kg})^{1/2} =$$
$$= 2.18 \cdot 10^6 \text{ m/s}.$$

Hence, the velocity of the electron on the ground state orbit in the hydrogen atom is 2200 kilometers per second. This is quite a velocity; it is 70 times higher than the orbital velocity of Earth, 40 times that of the Halley comet at points closest to the Sun, and 30 times that of our Voyager at its maximum. But it is still **137** times smaller than the velocity of light. In the "new physics", in which physical explanations do not exist ("Explanations *Yok*"), this 137 ratio is a "magic" number, named "fine structure constant".

16.03. Calculation of the Radius of the Hydrogen Atom

We can now return to the previously obtained formula of the energy balance,

$$m_e v^2/2 = ke^2/2R,$$

from which we determine the orbital radius R,

$$R = ke^2/m_e v^2.$$

Substituting in the formula for R the orbital velocity v on the ground state orbit, we will calculate the radius R_1 of this orbit. We may now substitute all the known values, to obtain,

$$R_1 = (9 \cdot 10^9 \text{ Nm}^2/\text{C}^2)(1.6 \cdot 10^{-19} \text{ C})^2 / (9.1 \cdot 10^{-31} \text{ kg})(2.18 \cdot 10^6 \text{ m})^2 =$$
$$= 5.3 \cdot 10^{-11} \text{ m},$$

or $R_1 = 0.53$ angstrom.

This radius, R_1=0.53 angstrom, is the radius of the electron orbit in the "ground state". R_1 is also the accepted value of the radius of the hydrogen atom in the ground state, or the "Bohr radius of the hydrogen atom". The R_1 value is an important parameter in atomic physics, as also is the 137 ratio of the velocity of light to the velocity of the electron on Bohr's orbit.

Note that as much as atoms of different elements differ from one another by their masses (which vary from 1 to over 260) and by the number of electrons (from 1 to over 100), their "normal" radii vary by a factor of 3 only, and remain in the range of an angstrom.

16.04. Deficiencies in Rutherford's Planetary Model of Atoms

The 1911 Rutherford model of the hydrogen atom suffered from two major deficiencies. First, it contradicted the well-established principle of Maxwell's theory that *any accelerated motion of an electrically charged body must cause vibrations and waves in the electromagnetic field*. A charged body must therefore radiate and lose energy into the field as long as it is accelerated.

The orbital electron is constantly changing the <u>direction</u> of its velocity toward the center of the orbit. This means that the velocity vector of the electron is being constantly increased by the addition of a velocity vector, directed toward the center of the orbit. Hence, the orbital motion is an accelerated motion, with a *centripetal* acceleration.

Thus, by Maxwell, and by what we know in physics until now, the orbital electron should lose its kinetic energy, while radiating energy into the electromagnetic field. The electron should therefore be "falling" on a spiral towards the nucleus, until the bitter end of collapsing with it. Thus, electron orbits, and atoms in whole could not be stable, and should collapse.

However we know well that atoms in our world are remarkably stable, thank goodness. Hence, something in our knowledge does not fit reality. It is not Maxwell's principle, which was found right in all cases of accelerated motion of charges, including radiation of accelerated **free** electrons. It is also not Rutherford's model or calculation, which was done in accordance with the best in physics. Hence, if nothing is wrong, then something must be missing!

We will show that the missing part is the interaction with the bound particles of the epola within the space of the atom. The bound epola particles create a de Broglie standing circular wave, enveloping the orbital electron. The orbital electron and its standing de Broglie wave represent a closed system, the electromagnetic energy of which is preserved or maintained within the system.

The second deficiency of Rutherford's model is its inability to account for the linear spectra of atoms. For example, the spectrum of hydrogen contains four prominent lines in the visible range: an orange line, of frequency f=457 THz, an azure line, of f=617 THz, a blue line, of f=691 THz, and a violet line, of f=731 THz. It was shown by J.J.Balmer in 1885, that the frequencies of each spectral line in the visible range of the spectrum of hydrogen can be obtained from an empirical formula, just by substituting an appropriate integer (e.g., 3 for the orange line, and 6 for the violet line).

The dependence of the frequency on an integer, thus also the dependence of the radiation energy on this integer, as well as the line spectra *per se*, persistently imply that atoms exist in profoundly discrete energy states only. This is ignored by the Rutherford model, which allows the electron to rotate on any orbit, of radii ranging from slightly above zero, and up to infinity, thus "allowing" the atom to absorb or emit any amount of energy.

16.05. Bohr's Orbits and Theory of the Hydrogen Atom

The theory of the hydrogen atom which agrees with the observed spectrum, was proposed by N.Bohr in 1913. In order to overcome the deficiencies of the Rutherford model, Bohr postulated that
> the atomic electrons may move only on certain allowed orbits;
> when the electron moves on an allowed orbit,
> > it does not radiate energy, and the orbit is stable.

Bohr also postulated that the radius R_n of the n-th allowed orbit can be calculated by multiplying the radius R_1 of the ground state orbit by the squared integer n. Thus,

$$R_n = n^2 \cdot R_1 .$$

Hence, the radii of the allowed orbits are

0.53 Å for the ground state or "first" orbit, n=1,
4 · 0.53 Å = 2.1 Å for the second orbit, n=2,
9 · 0.53 Å = 4.8 Å for the third orbit, n=3,
16 · 0.53 Å = 8.5 Å for the fourth orbit, n=4,

and so on.

As the radii of the allowed orbits <u>increase</u>, the orbital (kinetic or binding) energies E_n of the electron, <u>decrease</u> at the same rate, i.e.,

$$E_n = E_1/n^2 .$$

Here E_1 is the energy of the electron on the first or ground state orbit, equal to the 13.6 eV ionization energy of the atom. On the second orbit, the electron energy is

$$E_2 = 13.6 \text{ eV}/4 = 3.4 \text{ eV} .$$

On the third orbit, n=3, $E_3 = 13.6 \text{ eV}/9 = 1.5 \text{ eV}$;
on the fourth orbit, n=4, $E_4 = 13.6 \text{ eV}/16 = 0.85 \text{ eV}$; and so on.

For the frequencies and energies of radiation, emitted or absorbed by the atom, Bohr postulated that
> when the atomic electron makes a transition
> > from one allowed orbit, of energy E_n ,
> > to another allowed orbit, of energy $E_{n'}$,
> then the atom emits or absorbs a photon
> > of energy $E_p = hf$, equal to the difference

between the energies of the electron on these two orbits. Thus,

$$E_p = hf = E_n - E_{n'}.$$

With his theory, Bohr calculated all the then known lines in the hydrogen spectra as electron transitions between the allowed orbits. When techniques were later developed to investigate the far infrared part of the spectrum, the discovered new series of lines completely agreed with Bohr's theory.

In spite of this success, it is not right to see in Bohr's theory an explanation of the hydrogen atom and spectra. The theory is based on postulates, which were introduced without explanation. It remains unexplained, as it was before, why the centripetally accelerated orbital electron does not radiate its energy, and why the radii and orbital energies are as postulated.

16.06. Emission and Absorption of Light by Hydrogen Atoms

An atom can absorb energy in collisions with other atoms, (e.g., in thermal collisions) or with particles of nuclear matter, or from electromagnetic radiation, i.e., from vibrations of the bound particles of the electromagnetic (EMG) field. The absorbed energy may either increase the energy of the motion or position of the whole atom, or it may increase the energy of one or more of its orbital electrons. In this case the absorbed energy transfers the atom from its ground state into an "excited state", i.e., it transfers one or more orbital electrons to a more remote allowed orbit, of a higher n number.

The excited atom may pass to an excited state of lower excitation energy, or return to its ground state, by emitting the excitation energy into the EMG field. In the field, the emitted energy causes an electromagnetic wave, which spreads with the velocity of light c.

In the hydrogen atom with its one and only orbital electron, the excited states of the atom differ from one another by the radii of the electron orbit, i.e., by the energies of the orbital electron on these orbits. The largest energy which the atom (or orbital electron) can absorb is 13.6 eV. This energy ionizes the atom, i.e., takes the electron so far away, that the electron does not "feel" the attraction to the nucleus; 13.6 eV frees the electron out of the atom. For the free electron, its binding energy to the atom, and its orbital kinetic energy are both zero.

Now if the free electron were to fall back onto its ground state orbit, the atom would emit into the field the energy of 13.6 eV, i.e., it would create an electromagnetic wave (in the ultraviolet range), transferring this energy. In the "quantum" language we say that the atom emits a 13.6 eV photon. Now the binding energy of the electron to the atom, and the kinetic energy of the orbital motion of the electron, are equal to 13.6 eV each.

Let us now consider that the electron on its ground state orbit absorbed the energy of 10.2 eV. This would reduce its binding energy and the kinetic energy to

$$13.6 \text{ eV} - 10.2 \text{ eV} = 3.4 \text{ eV}.$$

The energy of 3.4 eV is a quarter of the binding energy of the electron on the ground state orbit. Hence, absorption of the energy of 10.2 eV increases fourfold the radius of the orbit, from 0.53 angstrom to 2.1 Å, i.e., the electron would transit to the allowed orbit #2, of n=2.

Returning from this excited orbit (#2) to the ground state orbit (#1), the electron would emit the energy of 10.2 eV into the field, creating an electromagnetic wave which transfers this energy. The wave is in the ultraviolet range; in quantum terms, the atom emitted a 10.2 eV ultraviolet photon.

The electromagnetic waves corresponding to all possible allowed transitions of the orbital electron appear as emission or absorption lines in the radiation spectra of hot hydrogen gas. They were observed, photographed, or otherwise detected and documented in various spectrometers by numerous researchers.

16.07. Combined "Bohr-De Broglie" Condition for Orbit Stability

We will show now that the physical explanations of the Bohr postulates is linked with the de Broglie waves of the orbital electron. The wavelength w_B of the de Broglie wave of an electron moving with velocity v is

$$w_B = h/m_e v .$$

Here h/m_e is Planck's constant, divided by the mass of the electron,
$$h/m_e = 6.63 \cdot 10^{-34} \text{ Js} / 9.1 \cdot 10^{-31} \text{ kg} = 7.3 \cdot 10^{-4} \text{ m}^2/\text{s}.$$
Thus,
$$w_B = (7.3 \cdot 10^{-4} \text{ m}^2/\text{s}) \cdot v ,$$
which is a handy formula to calculate the de Broglie waves of electrons. Substituting for v the velocity of the electron on the ground state orbit, we obtain
$$w_1 = (7.3 \cdot 10^{-4} \text{ m}^2/\text{s})/(2.2 \cdot 10^6 \text{ m/s}) =$$
$$= 3.3 \cdot 10^{-10} \text{ m} = 3.3 \text{ Å} .$$
The wavelength 3.3 angstrom is exactly equal to the circumference of the ground state orbit, which is
$$6.28 \text{ } R_1 = 6.28 \text{ } 0.53 \text{ Å} = 3.3 \text{ Å} .$$
Hence, the circumference of the ground state orbit contains **one** de Broglie wave, enveloping the orbital electron.

The velocity of the electron on the second, n=2, orbit is half its velocity on the ground state orbit, i.e., v_2=1.1 Mm/s. The wavelength w_2 is therefore twice that on the ground state orbit, w_2=6.6 Å.

The radius of the second allowed orbit is four times the radius of the ground state orbit, R_2=2.1 Å. The circumference of this orbit is
$$6.28 \text{ } R_2 = 6.28 \text{ } 2.1 \text{ Å} = 13.2 \text{ Å} ,$$
which is exactly twice the de Broglie wavelength w_2. Hence, the circumference of the second allowed orbit contains **two** de Broglie waves of the orbital electron.

Similar results are obtained on each allowed orbit. We conclude therefore that
> circumferences of allowed atomic orbits
>> must contain integral numbers of de Broglie waves
>>> of the orbital electron;
>> the number of the waves on each orbit
>>> is equal to the n number of the orbit.

This is the "combined Bohr-de Broglie" condition for orbit stability, which may replace Bohr's postulate for orbit stability. In this form, there is still no explanation why the centripetally accelerated electron does not radiate when circling on a stable orbit. However this formulation prepares the basis for a physical explanation.

16.08. Physical Explanation of the Stability of Atomic Orbits

The explanation can be given by considering the interaction between the moving electron and the bound electrons and positrons in the epola space. The moving electron slides apart the bound particles around itself, causing their vibrations. These vibrations create an electromagnetic wave around the moving electron, which envelopes the particle and accompanies its motion.

The frequency of the accompanying wave is proportional to the velocity of the particle, and the wavelength is given by the de Broglie formula. The accompanying wave propagates like a guided wave, in a narrow channel directed along the motion of the electron.

When an orbit contains an integral number of de Broglie waves of the orbital electron, the accompanying wave is circling along the orbit with the velocity of light, and at each circling the wave reaches every vibrating bound particle in the channel at the same phase as it did in all previous passages, i. e., giving to each particle the same command as it was giving in all previous circlings.

As a result, the bound particles vibrate each with its own amplitude. Some of them have constantly a zero amplitude, and are the nodes of the wave. Other particles, at quarter wavelength from the nodes, have maximum amplitudes, and are the antinodes or loops of the wave. Hence, the superposition of the circling waves creates a closed circular standing wave along the orbit. It is well-known that the energy of such a wave does not radiate outwards, but is confined to the wave.

The standing wave pattern is rotating along the orbit with the constant velocity of the orbital electron, which is contained inside one of the half-waves. For example, on the ground state orbit the velocity of the electron is 137 times smaller than the velocity of light. Hence, the accompanying wave circles the orbit 137 times during the time of one rotation of the orbital electron, enveloped in the standing wave pattern of the accompanying de Broglie wave.

It is important to remember that both the accompanying wave and the standing wave are due to the motion of the orbital electron, and to the vibrations of the bound particles in space. As in any wave motion, the particles of the ambient which carries the wave, in this case the bound electrons and positrons of space, do not move with the wave. They remain around their equilibrium positions, and the amplitude of their vibrations is much smaller than the 4.4 fm distance between the particles, and millions of times smaller than the radius of the orbit.

The rotation of the orbital electron with constant velocity inside one of the half-waves of the closed circular standing-wave pattern is mechanically analogous to the rotation of a balanced, wave-shaped ring or hoop (or tire). The centripetal accelerations of the particles of such a rotating ring are provided by the internal bonds between the particles of the ring, and are a matter of their "internal concern". In the rotation of the ring (or tire) as a whole, there is no centripetal acceleration at all!

Hence, the orbital electron, as a particle of a rotating ring, obtains its centripetal acceleration in its interaction with the surrounding particles of the ring. With respect to the electromagnetic field outside the ring, the electron is not centripetally accelerated, therefore it does not radiate its rotational energy into the field.

If an electron happens to move on an orbit with a velocity, which does not create an integral number of de Broglie waves on this orbit, then a closed standing wave pattern cannot be formed. At each rotation of the accompanying wave along such an orbit, every bound particle in the channel of the wave is reached by a phase which differs from the phase (or command) given to it at the previous passage. Therefore, the "rotating ring" wave pattern is not formed, and electromagnetic energy is not preserved within the channel of the accompanying wave. The electron is centripetally accelerated with regard to the outside field, and radiates (loses) energy into it. Such an orbit is not stable.

16.09. Spectra and Stability of Atoms Heavier Than Hydrogen

The Bohr theory, so successful for the hydrogen atom, was unable to account for the spectra of other atoms, not even for the spectrum of helium. First of all, it was clear that the electron orbits in atoms having two or more orbital electrons cannot be considered as circles but should be elliptic. An elliptic orbit has two focal points, and the atomic nucleus resides in one of them. The velocity of the electron on an elliptic orbit is not constant: on the parts of the ellipse which are closer to the nucleus the electron moves faster.

The problem of planetal motion on elliptic orbits was solved by J.Kepler in 1609. However the motion of a <u>charged</u> body on an elliptic orbit is complicated by the inevitable energy losses to the electromagnetic field, when the velocity of the body is changing. Therefore, an atomic electron moving on an elliptic orbit should lose energy to the EMG field and "fall" towards the nucleus, even if the mean radius of the orbit (i.e., half the sum of the short and long axes of the ellipse) is "allowed" by Bohr's theory. Hence, orbits of atoms, and atoms themselves should be unstable. But atoms <u>are</u> stable! Once again, based on well founded and working physical laws, we come to a conclusion which contradicts reality.

The contradiction can be relieved through consideration given to the epola structure of the EMG field, and to its interaction with the atomic electrons. However with the a priori postulated emptiness of space, the procedure in physics was to introduce new unexplained postulates and gimmicks which would allow mathematical derivations and calculations. This was the way (and still is) of the then created (1924 - 1926) new science of quantum mechanics.

To deal with atomic structure and spectra, quantum mechanics accepted Planck's postulate of the quantization of radiation, and Bohr's postulate on the stability of atoms having orbital radii and energies expressed through his integer n. This integer became the "main quantum number" of the atomic electron.

In addition to Bohr's postulate, it was postulated that
    an electron does not radiate
        while accelerating on an elliptic orbit,
  if the ratio of the axes of the ellipse
    has certain quantized values, expressed by
the integer l_q, named "orbital quantum number".

This postulate takes care of the contradicting radiation of the accelerated electron, in the same way as done by Bohr's postulate. Observed spectral lines are considered as electron transitions between stable orbits, and are used to define the quantum conditions for them. These conditions are then proclaimed as conditions for orbit stability, i.e., conditions under which there is no radiation of the accelerated orbital electron. And no explanation is given (or looked for) of why it is the way it is.

A third quantum number, m_q, named "momentum quantum number" was introduced to express the allowed quantized positions of the plane of the orbit relative to the planes of the other electron orbits in the atom. Finally, the "spin quantum number" s_q was introduced to account for the two possible values of the spin of the orbital electron: +1/2, for "spin up", and -1/2 for "spin down" position.

The four quantum numbers

$$n, \; l_q, \; m_q, \; \text{and} \; s_q,$$

fully determine the energy and momentum state of an orbital electron in the atom. Any amount of energy, emitted or absorbed by an atom, i.e., any spectral line in atomic spectra, is always due to a transition of an orbital electron from an orbit, identified by these four quantum numbers, to an orbit in which at least one quantum number has a different allowed value.

16.10. Superposition of Waves

The de Broglie wavelength of the accompanying wave of the orbital electron in the ground state of the hydrogen atom, calculated in Section 7, is 3.3 angstrom, equal to the circumference of the orbit and of the atom. Therefore, the accompanying wave of the orbital electron involves all bound electrons and positrons of space, which happen to be in the peripheral shell in and around the atom.

This means that the same bound particles of the epola space must simultaneously participate in the accompanying waves of all orbital electrons of the atom. Hence, the accompanying waves of the orbital electrons are SUPERPOSED on one another. However, the superposed waves do not necessarily merge or interfere with one another. As a matter of fact, the conditions for a merge or for interference are so selective, that very seldom can these phenomena occur when different waves are superposed.

It is a well-known fact that a particle of a medium can simultaneously perform many vibrations, in different directions, with different frequencies. As long as the directions or the frequencies of these *superposed* vibrations differ from one another, the vibrations do not merge or interfere; each vibration has its individual existence, and lasts until its energy is dissipated or randomized in the medium.

It is also a well and long known fact that any spot of a medium can simultaneously carry many waves, of different frequencies, different directions (and velocities) of propagation, and of different MODES or POLARIZATIONS, i.e., of different directions and planes of particle vibrations. As long as the *superposed* waves differ by at least one of these features, the waves do not merge or interfere. Each wave has its own separate existence, and lasts until its energy is either absorbed by a body or dissipated in the medium.

This is the content of the SUPERPOSITION PRINCIPLE in physics, stating that differing superposed waves do not merge or interfere, no matter how densely "packed". For example, there is no interference or merging of the thousands of different sound waves, emitted by a great symphony orchestra, though all these waves propagate simultaneously and fill up the concert hall. Moreover, all these sound waves, of wavelengths ranging from 10 meters to 5 centimeters, simultaneously elbow their way through the centimeter-wide opening in our ears, and through the even narrower auditory canal in them, but there is neither merging nor interference of the *superposed* (and "supersqueezed") waves there.

Similarly, thousands of different light-waves, from all spots of viewed objects, simultaneously squeeze their way through the pupils of our eyes, and through the diaphragms of cameras, but there is neither merging nor interference of these superposed waves there. The perceived "merging" into "white" light, of waves, individually perceived as red, yellow, green, blue, etc., spectral "colors", occurs in the retina, and is a biochemical and neurological process, not a physical merging of electromagnetic waves.

16.11. Merging and Interference of Waves

To merge or interfere, superposed waves must be COHERENT. Usually, to be coherent, the waves have to be emitted at the same time by the same source. Turning other waves into coherency is a highly complicated technological enterprise. Then, the coherent waves, brought together or superposed after having crossed different paths in different media, must have identical frequencies.

In addition, the directions and planes of vibrations of particles in the waves, i.e., the *modes* or *polarizations* must either be identical or correspond to each other. Coherent waves of identical frequencies but of different modes or polarizations, may in some media propagate with differing velocities. This alone may prevent merging or interference of such superposed waves.

The severe conditions for merging or interference of superposed acoustic waves can be fully understood only if the "continuous media" model, with all its mathematical advantages, is given up, and the waves are considered as being carried by the constituent particles of the atomic bodies, i.e., by the molecules, atoms, or ions. Hence, each wave should be considered as consisting of more or less cooperating vibrations of individual constituent particles of the medium, with their individual motions, and with their binding to the neighbors.

The similar severeness of the conditions for merging or interference of superposed electromagnetic waves leads to the idea that these waves, too, consist of more or less cooperating vibrations of individual particles, existing in space. Hence, trials to give a physical explanation of merging and interference conditions, and of the superposition principle, also lead to the conclusion that space and the electromagnetic field must contain bound particles, and cannot be continuous or empty.

In this case, too, the quest for understanding is dangerous to the postulated emptiness of space. Therefore, new physics is not only opposed to the search for physical explanations concerning the interference and superposition of electromagnetic waves, but is also reluctant and suppressive to existing century-old explanations of these phenomena in acoustics. The reason here is that the physical atomistic model provides understanding, but the needed calculated results are provided by the continuous media model. Mathematics is still unable to provide calculated results in a wave mechanics, which is based on atomistic presentations.

16.12. The Pauli Exclusion Principle

From the analysis of atomic spectra, W.Pauli derived in 1925 his "exclusion principle", stating that
> no two electrons in an atom
> can exist in the same energy state.

In other words, if the energy and momentum of an orbital electron in an atom are defined by a certain set of values of the four quantum numbers,

$$n, \; l_q, \; m_q, \; \text{and} \; s_q,$$

then there cannot be in the same atom another orbital electron having the same values of all the four quantum numbers.

The exclusion principle plays an important role in physics. This includes not only atomic spectroscopy but also atomic structure, the determination of electron configurations in atoms, and the periodic classification of the chemical elements and isotopes.

The exclusion principle was also generalized to include electrons confined to proximity in space. For example, if in a solid body the distance between the 'free' electrons is ~1 nanometer or less, then in this body (usually metal) there cannot be two electrons having the same energy and spin. Hence, the exclusion principle became a pillar of the electron theory of metals and of solid state physics on the whole.

In quantum theory, the Pauli exclusion principle is presented as a kind of unexplainable "divine revelation". Seeking explanations for this principle, as well as of any one of the quantum postulates, is considered an apostasy. However the electron-positron structure of the EMG field allows quite a simple physical explanation of this postulated principle.

The de Broglie wavelength of the accompanying wave of the orbital electron in the ground state of the hydrogen atom, calculated in Section 7, is 3.3 angstrom, equal to the circumference of the orbit and of the atom. Therefore, the accompanying wave of the orbital electron involves all bound electrons and positrons of space, which happen to be in the peripheral shell in and around the atom.

This means that the same bound particles of the epola space must simultaneously participate in the accompanying waves of all orbital electrons of the atom. Hence, the accompanying waves of the orbital electrons are SUPERPOSED on one another. However, as long as they differ from each other by an energy, characterized by at least one quantum number, the waves do not interfere with one another.

The numerous orbital electrons in an atom may approach each other, or their orbits may cross. Even if such an event occurs, the "close encounter" would last for an extremely short time, because of the very high orbital speeds, and because the orbits diverge. If two electrons are exposed to their mutual electrostatic repulsion during the encounter, then, due to the shortness of the interaction time, no changes in their motions succeed in occurring.

If there were in the atom two electrons with identical four values of their quantum numbers, then their two accompanying waves would be identical. But two identical waves, carried by the same bound particles, would very soon merge into one wave. This means that the two electrons would then be moving with identical velocity on the same orbit, and would be exposed to a lasting electrostatic repulsion between their charges.

Obviously, even before the two orbits merge, the electrostatic repulsion becomes sufficiently strong to flip over the spin of one of the electrons, or to force the orbital plane of an electron into another allowed position, or to otherwise change one or more of the quantum numbers of the electrons. This explains the Pauli principle.

16.13. On Achievements and Misguidings Of Quantum Mechanics

Quantum mechanics was from its beginning a very sophisticated mathematical theory and, in the years of its successful march through physics, it became more and more sophisticated. Quantum mechanics can account for the complexities of atomic spectra, for the structure of atoms, molecular binding, crystal structure, and so on. There is not a single branch of physics and chemistry which did not gain from the "invasion" of quantum theory into it.

However, the quantum theory did not contribute towards the understanding of physical phenomena. For each difficulty, it introduces *ad hoc* postulates, principles, forbiddance, and exclusions, without (and against) trials of explanation. Moreover, the inability of this theory to explain physical phenomena is dismissed by postulating that there are no explanations, and that all there is and can be, are the mathematical equations and their solutions. As a result, quantum mechanics itself is not understandable. R. Feynman expressed it in 1967, by saying: "I think I can safely say that nobody understands quantum mechanics".

Being a mathematical science, not based on a physical model, and not aimed to develop a physical model, quantum mechanics is unable to provide physical explanations for its principles and findings. This is not a misdeed, just unfortunate. If the founders of this science were to accept this inability as unfortunate, to express their sorrow about it, together with a hope for a better future, then this could have only increased our gratitude to them. After all, if you ask for a product and get in reply: "Sorry, we are temporarily out of it, please check again", then you can only thank the salesman.

However the creators, apostles, and apparatchiks of quantum theory present their inability to explain as an achievement to be proud of, as a sign of the divinity of their postulations. They insist that there are no explanations and cannot be, that all there is are their principles and equations. They deny physical causality and the reality of atomic orbits, and consider anyone seeking understanding as an ignoramus.

Similarly, a merchant asked for a product which he does not have, would say:
    "there is no such thing"; (Malta *Yok*, Explanations *Yok*);
    "you will never find it"; (*Ignorabimus*);
    "they quit making it"; (Explanations *non fingo*).
If you believe him, then he is right, polite, and blown up with self-dignity and importance. But if you show signs of disbelief, then he announces that "you don't know what you are talking about" (or something worse). Shop around, and you may, sometimes amazingly easy, find the merchandise in another store.

CIRCLE SIX

Chapter 17
VELOCITIES OF ATOMIC BODIES AND OF NUCLEAR PARTICLES IN THE EPOLA

17.01. Velocity Limits of Atomic Bodies in the Epola Space

In a resting atom, there is an epola deformation cluster around the nucleus, and there are epola waves, accompanying the motion of each orbital electron. When the atom <u>moves</u> in the epola, then an epola wave is accompanying the motion of the nucleus, and epola accompanying waves, corresponding to this motion, are added to the accompanying waves of each orbital electron.

As long as the velocity of the atom is much smaller than the velocities of the orbital electrons, the addition of the accompanying waves does not affect the structure and the stability of the atom. For example, the velocity of Earth in its motion around the Sun is 30 km/s, which is 73 times smaller than the orbital velocity of the electron in the hydrogen atom. Hence, all atoms of and on earth easily adjust to this motion.

The accompanying waves of each nucleus, and the additional accompanying waves of each atomic electron, cause the epola units to open and close coherently for the motion of these particles. Hence, the added accompanying waves make the epola transparent for the motion of atoms and atomic bodies.

But if and when the velocity of the atom becomes equal or larger than the orbital velocity of an orbital electron, then the stability of this electron on its orbit might become fluttery. With increasing velocity, the probability increases that the atom might lose its outermost electron(s), and turn into an ion. One would then observe the ionization of atoms due to their fast motion.

We should thus consider the 2200 km/s velocity of the "ground state" electron in the hydrogen atom as the threshold velocity of any atomic body relative to **our** surrounding epola. At such a velocity, the molecular bonds between atoms in the body are likely to weaken or disappear, and the body might disintegrate into separate atoms and ions.

However, we are secure from reaching such velocities, because no atomic body in the Solar System and vicinity comes close to them. The velocity of the fastest planet (Mercury) does not exceed 50 km/s, the highest velocity of comets is in the range of 60 km/s, and such also was the velocity of our Voyager. The velocity of the manned Apollo mission to the Moon was only 10 km/s.

The danger level for human travel can be set at a velocity of ~100 km/s. At this velocity, the added energies of the accompanying waves of the nuclear particles, of which the atoms of our bodies consist, may raise the temperature of our bodies to 40°Celsius (105°F). The velocity limit of 100 km/s for human travel allows us to reach the Moon in slightly above an hour, to reach Mars in 10 days, and Pluto (end of the Solar System), in 2 years. However this velocity limit absolutely excludes the completion of even the shortest interstellar tour during a human lifetime, however lengthened by science and technology.

17.02. Einstein's Formula for the Dependence of Mass on Velocity

Based on J.C.Maxwell's theory of the electromagnetic field, and using the mathematical Lorentz Transforms, Einstein made in 1905 his multipage derivation of the dependence of the mass of an electron on the velocity of its motion in the field. In his formula, the increased mass m_v of an electron, moving in the electromagnetic field with a velocity v, is

$$m_v = m_e(1 - v^2/c^2)^{-1/2} .$$

Let us calculate, e.g., the increased mass m_v of the orbital electron in the hydrogen atom. The velocity v of this electron is 137 times smaller than the velocity of light c. Hence the ratio v^2/c^2, appearing in the formula, is $1/137^2$, or $5.3 \cdot 10^{-5}$. Substituting this value into the formula, we obtain,

$$m_v = m_e (1 - 5.3 \cdot 10^{-5})^{-1/2} = m_e (1.000027).$$

Hence, the mass m_v is only by ~0.0027%, or by less than three thousandths of a percent larger than the mass m_e of the electron. A hundred-fold increase of the velocity of the electron, to 220,000 km/s, would bring the v^2/c^2 ratio to 0.53. Substituting this value to the formula yields

$$m_v = m_e(1 - 0.53)^{-1/2} = m_e \cdot 1.46.$$

Hence, at this velocity, the mass m_v is by 46% larger than the mass of the electron. With a further increase of the velocity of the electron, the mass m_v increases drastically, and at $v=c$, the formula yields for m_v the value of "infinity".

An infinitely large mass of an electron would mean that no matter how large a force acts on such an electron, the velocity of the electron does not increase, and the acceleration of the electron is zero. Therefore, believing in the absolute power of his formula, Einstein stated that "*Velocities greater than that of light ... have no possibility of existence.*"

Einstein's denial of the possibility of "superlumic" motion is a result of an improper extrapolation of the power of his formula beyond the limits of its applicability. We know now that electrons and some other nuclear particles can be accelerated to superlumic velocities. Such "superlumic" particles, and the observed phenomena caused by super-lumic motions are named "super-relativistic", or "ultra-relativistic", in order not to openly denounce Einstein's statement.

Einstein improperly generalized his results, derived for electrons only, onto atomic bodies, by saying that his expressions "*must also...apply to ponderable masses as well.*", and that "*these results as to the mass are also valid for ponderable material points, because a ponderable material point can be made into an electron ... by the addition of an electric charge, no matter how small.*"

This automatic generalization is physically improper, for it does not consider the tremendous differences between electrons and atomic matter, particularly, the quadrillion times larger density of matter in the electron, and the 30 orders of magnitude larger density of electric charge. The generalization makes one believe that atomic bodies can be accelerated, just like electrons, to velocities close to the velocity of light, while we know, that at hundred time smaller velocities, atomic bodies may disintegrate, turning into separate atoms and ions.

17.03. Effects of Einstein's Dismissal of Newton's Concept of Mass

Newton introduced the concept of the mass of a body as of the quantity of matter in it. The Newtonian mass is indestructible and uncreatable, and also unchangeable, for as long as the body remains unchanged. This was proven by generations of physicists and chemists, and is included in Lavoisier's Law of Mass Conservation.

Hence, perhaps the worst thing which Einstein the mathematician (not the great physicist, which he was, too) did to the natural science of physics, was his dismissal of the Newtonian concept of mass, by concluding from equations, that the mass depends on velocity, increasing up to infinity, and that mass and energy are mutually interchangeable, turning into one another.

The ease of giving up fundamental concepts in physics for the sake of "simpler" interpretations of mathematical formulae and experimental results, reminds one of a student, who found that the accelerations of a wooden block, sliding along an inclined board, are not the same on different segments of the board.

The student was educated to believe that his "creative thinking" is above all. He was thought "group theory" before elements of arithmetic, and relativity before elements of physics. He thus decided, that what is good for Einstein is also good for him, and interpreted the results of his experiment as due to the changing mass of the block. He also stated, that on a certain segment, where the block stopped completely, its mass just became infinitely large.

When asked if he shouldn't rather consider the changing <u>resistance</u> to the motion on different segments of the board, the student replied: "I don't care! Newton's Law of Motion states that the acceleration of a body is equal to the force divided by the mass. If the gravitational force acts, but the acceleration is zero, then the mass of the body must be infinitely large". Amazingly, other students quietly accepted this "high science" argumentation.

We may thus find that to people "affected" by mathematical thinking, it is easier to stick to a formula, even if it leads to infinite masses, than to consider the interaction with the ambients, and the additional forces, resulting from it.

The "I don't care" approach is not just a game; it is often used in very serious physics, too. The approach was adapted in Solid State Physics, in the measurements and calculations of the apparent masses of the 'free' or conduction electrons in bodies.

The values of the apparent masses of 'free' electrons in solids follow mostly from the measurements of electric currents in the bodies, and strongly depend on external and internal conditions. Hence, the so-measured masses of the conduction electrons in solid may be many times larger or smaller than the mass of the electron. The apparent mass may also be negative, meaning that the electron in the body is accelerated in a direction, opposite to the force of the external electric field. However, unlike the relativists, solid state physicists distinguish the unchangeable mass of the electron from these alterable apparent values, by naming them "effective masses".

17.04. Physical Reasons for the Dependence of Mass on Velocity

We found that the motion of a free nuclear particle in the epola, with a velocity v much smaller than the velocity of light c, creates around the particle an epola wave, which accompanies the motion. The frequency of the wave is proportional to the velocity of the particle, and the wavelength is equal to the wavelength of the de Broglie "wave of matter" for this particle.

As do all epola waves, the accompanying wave propagates with the velocity of light c, while the accompanied free particle moves much slower. Hence, the accompanying epola wave succeeds to undergo multiple reflections from the epola in front of the moving particle, behind it, and sidewards.

Thus, a narrow channel is created in the epola around the particle, inside which the multiply reflected accompanying wave forms a *standing wave pattern*, like in a *wave guide*. The length of the wave-guide channel is several wavelengths in front of the particle, and the width is smaller than half a wavelength. The relations of these sizes to the wavelength decrease with increasing velocity of the particle.

The standing wave pattern is enveloping the particle, and is moving with the velocity of the particle (not of the wave!). Within this standing wave pattern, the epola is vibrationally pre-formed for the motion of the particle. This means, that each epola unit, which the particle is about to enter, "opens" for the particle, and "closes" when the particle is about to leave. The coherence of the accompanying wave with the *established* motion of the free particle excludes the probability of collisions between the free particle and the bound particles of the epola. Hence, the epola becomes VACUUM-TRANSPARENT for the motion, as transparent as a <u>hypothetical absolute emptiness</u> would be.

The epola vibrations, sustained by the established and coherent accompanying wave, turn the wave also into an <u>active</u> participant of the motion of the free particle. For when the bound particles of an epola unit move apart in front of the free particle, the unit exerts a "soaking in" action, which enhances the forwarding of the particle, and "refreshes" the particle's "memory" of the motion, in case it "hesitates" to enter. When the particle is about to leave the epola unit, the convening motions of the bound particles of the unit exert a "squeeze out" action on the free particle, in case it "hesitates" to leave.

This active participation of the accompanying wave in the preservation of the established motion of the particle is also the physical reason for inertia. Hence, inertia and the "inertial mass", are features not only of the moving particle, but of the system "moving particle and its accompanying wave". When something would tend to reduce the established velocity of the particle, then energy would be "pumped" from the wave to the particle. At a tendency to increase the established velocity, the wave would "soak" the extra energy. Hence, the accompanying wave is well rewarding the motion for the energy which was invested into it, "by inertia", in the process of establishing the motion.

When the particle is *sublumic*, i.e., its velocity v is close to the velocity of light c, then the accompanying wave may not have sufficient time to pre-form the epola for the motion. The moving free particle must then by itself shift apart the epola particles on its way; it must "open the gates" to and through epola lattice units "with its own body". Therefore, the motion faces resistance, which sharply increases, as the velocity of the free particle becomes closer to the velocity of light.

The appearance of a resistance to the motion, and the increase of the resistance with increasing velocity, can be detected and measured as a decrease in acceleration, submitted to the free particle by the action of a constant accelerating force. Considering that the acceleration is equal to the force, divided by the mass, and finding that the acceleration is decreasing, though the force remains constant, the "I don't care" mathematician concludes that the mass of the body is increasing.

Increasing is, to us, the *effective mass* of the <u>system</u>, consisting of the moving particle and of its accompanying wave. The mass of the moving particle remains constant at all times. However, the quasi-mass of the photon of the accompanying wave is increasing with increasing velocity, and this increase depicts the fact that with increasing velocity the wave has less and less time to appropriately shift apart the epola particles. The shorter the time the harder it is to perform the shift, the more energy of the accelerator must go into the wave.

We shall try our model on the case of a free electron, of mass m_e, moving with a sublumic velocity v. The observed effective mass of the moving system is then m_v, and consists of the real mass m_e of the electron and of the unknown quasi-mass m_B of the photon of the accompanying wave. Hence, our task is now to determine m_B.

17.05. Formulae for the Photon Energy, Momentum, and Mass in Accompanying Waves

The wavelength w of the accompanying wave of an electron, moving in the epola with a velocity v, is given by the formula,

$$w = h/m_v v .$$

This is also the de Broglie formula for the wavelength of his "waves of matter". Here h is the Planck constant,

$$h = 6.6 \cdot 10^{-34} \text{ J·s},$$

and m_v is the "effective" mass of the moving electron, increased because of its fast sublumic motion.

Let us now multiply both sides of the equation by the frequency f of the accompanying wave, to obtain,

$$fw = hf/m_v \cdot v \ .$$

The frequency f is equal to the velocity c of the wave, divided by the wavelength w. On the left-hand side of the equation, the value f, equal to c/w, is multiplied by w. Hence, the left-hand side of the equation is just the velocity of light c,

$$fw = (c/w)\,w = c \ .$$

On the right-hand side of the equation, the numerator is hf, which is, by Planck's Law, the photon energy E_p of the wave, E_p=hf. Therefore, our equation becomes

$$c = E_p/m_v \cdot v \ ,$$

and the photon energy of the accompanying wave is

$$E_p = m_v \cdot v\,c \ .$$

The momentum of a particle, thus also of the quasi-particle photon, is equal to the energy of the particle, divided by its velocity. The velocity of the photon is c, therefore the momentum of the photon of the accompanying wave is

$$E_p/c = m_v \cdot v \ .$$

The mass of a particle is equal to the momentum of the particle divided by its velocity. Hence also the mass m_B of the quasi-particle photon of the de Broglie wave (or of the accompanying wave) is

$$m_B = m_v \cdot v/c \ .$$

One should remember that the energy of the photon is actually the energy transferred in the wave from one bound particle of the epola to the next in row. When there is a target particle in the way of this energy transfer process, then the target may either absorb the photon energy, or not. To the target particle, the energy of the photon is actually the energy transferred (or not) to the target by the last epola particle in row.

Similarly, the momentum of the photon is actually the momentum transferred in the wave from one bound epola particle to the next in line. When there is a target particle in the way of this transfer process, then the target may either absorb the momentum of the photon, or not. To the target particle, the momentum of the photon is just the momentum transferred (or not) from the last bound epola particle in the line.

In a short slogan-like statement, we may say that
> behind every quasi-particle photon,
> of a given frequency, energy, and momentum,
> there is a real bound electron and/or positron of the epola,
> which vibrates with the frequency of the photon,
> and transfers to the next epola particle,
>> or to a target particle,

the energy and momentum, considered to be those of the photon.

17.06. The Effective Mass and Momenta in the Sublumic Motion of a Free Electron In the Epola

We consider the effective mass m_v of a free electron, moving in the epola with a velocity v, as being the mass of the system "moving electron and its accompanying wave". Hence, m_v consists of the constant mass m_e of the electron, and of the quasi-mass m_B of the photon of the accompanying wave of the electron. In a short notation,

$$m_v = m_e \leftrightarrow m_B .$$

The wavy plus sign indicates that we do not know how to sum the stable mass m_e with the quasi-mass of the photon, transferred among the epola particles with the velocity of light. Hence, the sign is meanwhile inoperable. Substituting for m_B the value,

$$m_B = m_v \cdot v/c,$$

found in Section 5, yields

$$m_v = m_e \leftrightarrow m_v \, v/c .$$

Multiplying both sides of this equation by c, we obtain

$$m_v c = m_e c \leftrightarrow m_v v .$$

The equation for masses became an equation of momenta, in which:

$m_v \cdot c$ is the momentum of the system *electron and accompanying wave*
    having an effective mass m_v, and
    moving in the epola with the "*lumic*" velocity c ;

$m_e \cdot c$ is the momentum of a *lumic* electron, of mass m_e,
    moving with velocity c in a hypothetical empty space,
    in which there is no epola, just nothing but geometry,
    thus no resistance to lumic motion; and,

$m_v \cdot v$ is the momentum of the photon of the epola wave, created
    when the lumic electron reaches a "gate" to an epola unit.

The radius of the electron or positron is ~50 times smaller than the lattice constant of the epola. Hence, the free area of the "gate" to an epola unit, between the bound particles, is ~2000 times larger than the cross-section area of a free or bound electron or positron. Therefore, an electron moving toward the gate with lumic velocity, i.e., without the pre-forming action of an accompanying wave, still has a very small probability (around 1/2000) of a head-on collision with a bound particle <u>at the gate</u>.

At the moment when the lumic electron arrives from the hypothetical Newtonian Empty Space to a "gate" of an epola unit, it pushes apart the epola particles there, thus initiating the accompanying wave. The epola particles, initiating the wave are pushed apart in a direction, perpendicular to the direction of the velocity and momentum of the lumic electron. Therefore, the momentum of the photon is <u>*at this instant*</u> perpendicular to the momentum of the lumic electron. We may therefore add the two momenta by the addition rules of any two "regular" mutually perpendicular vectors.

17.07. Calculation of the Effective Mass

By the law of vector addition, the sum of two mutually perpendicular vectors is represented by the hypotenuse of a right triangle, the two sides of which (cathetuses) represent the addend vectors. Recalling from our early encounter with geometry the Pythagoras Theorem (*the square of the hypothenuse is equal to the sum*

of the squares of the cathetuses), and applying it to the momenta in Section 6, we have
$$m_v^2 \cdot c^2 = m_e^2 \cdot c^2 + m_v^2 \cdot v^2 .$$
To find m_v from this equation (to solve it for m_v), we first subtract from both sides of the equation the term $m_v^2 \cdot v^2$, which yields
$$m_v^2 \cdot c^2 - m_v^2 \cdot v^2 = m_e^2 \cdot c^2 .$$
Then we group,
$$m_v^2 (c^2 - v^2) = m_e^2 \cdot c^2 ,$$
and divide both sides by $(c^2 - v^2)$, to get
$$m_v^2 = m_e^2 \cdot c^2 / (c^2 - v^2) .$$
In the right-hand side of the equation, we now divide the numerator and denominator by c^2, getting
$$m_v^2 = m_e^2 / (1 - v^2/c^2) .$$
Finally, we extract the square root of both sides of the equation, to obtain,
$$m_v = m_e / (1 - v^2/c^2)^{1/2} ,$$
or
$$m_v = m_e (1 - v^2/c^2)^{-1/2} .$$
This result is identical with Einstein's equation, presented in Section 2. Our formula is derived, based on physical assumptions, following from the epola structure of space, while using first year high-school algebra. Our way is in strong contrast with Einstein's multipage derivation, and with the complicated mathematical approaches which he had to employ. The ease and brevity of our derivation, accessible to high-school freshmen, should convince one that

> *based on the right physical model, one can do more*
> *and in simpler ways and forms,*
> *than when confined to mathematical models only.*

As are all physical laws, this formula is approximate, too. No matter how careful we are in establishing a law of nature, whether from observations, experiments, calculations, or all together, there are always some factors which were not or could not be taken into account. Under certain conditions, these usually negligible or non-influential factors or participants may become decisive, turning down some or all conclusions based on the law.

17.08. Lumic and Superlumic Motion of Nuclear Particles

Substituting v=c in Einstein's formula for the "dependence of mass on velocity", turns $(1-v^2/c^2)$ into zero, so that the denominator is zero, and the mass m_v becomes infinite. This made Einstein state that *"Velocities greater than those of light ... have no possibility of existence."*

However, it is not so. The <u>real</u> mass does not become infinite, lumic and superlumic motions of nuclear particles are possible, and have been achieved and observed, especially in the outcome of nuclear processes. The calculated turning of m_v to infinity expresses not an increase of the mass of the lumic or superlumic particle itself, but of the increasingly large amounts of energy, which have to be put into the particle, in order to enable the motion.

Lumic and superlumic particles have no accompanying waves to prevent collisions with the bound particles of the epola. Therefore, the particles lose energy in collisions, and quite soon their velocities decrease well below the lumic limit, when they can create the accompanying wave. Hence, superlumic motion of single particles of nuclear matter in the epola is possible, but cannot last long.

<u>Beams</u> of superlumic particles create breakthrough channels in the epola, analogous to breakthrough channels created by a lightning in the air and in insulating bodies: bricks, marble, etc. Within the break-through channel, the epola is "evaporated"; there is no lattice there but separate electrons, positrons, and electron-positron pairs, much farther apart than in the lattice. Hence, there are fewer collision events, and the energy losses of those particles of the beam, which were impoverished in a collision, are promptly replenished by the other particles of the beam, and by their (usually nuclear) source of energy.

One may wonder how it is that on the basis of the electron-positron lattice structure of space we derived the same formula as Einstein, who proclaimed the absolute emptiness of space. But let us not forget, that Einstein used the Maxwell Equations, and that these equations were based on Faraday's and Maxwell's electromagnetic field <u>in the ether</u>, i.e., in a quasi-material ambient, not in an emptiness.

Maxwell's electromagnetic waves are waves IN and OF the ether, and his displacement currents are currents due to the displacements of electric charges, again, IN and OF the ether. Using the Maxwell equations as the basis of his derivation, Einstein took with them also their hidden contents, namely the common physical resistance of the material, for which they were developed, to motions with velocities close to the wave velocities in this material.

With all our objections against the ether, considered as a continuous, elastic but massless medium, hence physically unfeasible, the ether concept was one of the pillars which enabled the relatively greatest achievements and progress of physics. Anyone who bases his derivations on the Maxwell Equations, or on other achievements of that period, may be introducing elements of a material base in his results, in spite of his devotion to the absolute emptiness of space.

Chapter 18

FREQUENCY EFFECTS AND VELOCITIES IN WAVE PROPAGATION

18.01. Wave Intensity and Quantum Energy

When a wave passes from one medium (*refracts*) into another one, in which the propagation velocity is different, then the direction of propagation changes, except in the case when the incident beam is perpendicular to the boundary between the media. In any case, some energy of the beam is *reflected* back into the first medium, and some is absorbed on the boundary. Hence, the energy, or intensity of the *refracted* beam is always smaller than the energy or intensity of the incident beam.

This "intensity energy" of the waves or beams is the sum of the vibrational energies of the particles of the medium, which vibrate in the wave. The vibrational energy of each particle is proportional to the square of the amplitude. Hence it differs from the quantum energy (photon or phonon energy), which is the energy transferred to the next particle in the beam, and is proportional to the wave frequency.

A reduction in the *intensity energy* or intensity of the wave, due to the spreading of the wave motion, to the splitting of the beams, or to absorption, is connected with the reduction of the vibrational energy of the particles of the medium, thus with the reduction of the amplitude of their vibrations. In these processes, in reflection, refraction, in propagation with spreading and absorption, when the wave intensities, energies, and vibrational amplitudes of particles decrease, the wave frequency usually remains invariant. Hence the quantum energy of the wave remains constant, too. However the number of photons in the beams is decreasing with decreasing intensity. The observed constancy of the wave frequency led to the formulation of the *principle of frequency invariance*.

18.02. The Principle of Frequency Invariance

The physical reason for frequency invariance is that in an established wave motion, there are quadrillions and quintillions of quintillions of particles vibrating with the frequency of the wave. When the wave, involving such large amounts of vibrating particles, reaches a boundary with another medium or region, then these vibrating particles force the particles in the other medium or region to vibrate with the same frequency. Hence, the frequency of the incident wave is transferred intact to the refracted wave but not the vibrational amplitude and intensity.

Any factor, tending to change the frequency of an established wave motion, faces the strong resistance of the quintillions of particles, vibrating in the wave. A factor, which is not sufficiently strong or "serious", will soon "find out", that it cannot "beat them", and will be forced to just join their frequency.

Hence, the physical reasons for frequency invariance are the same as for inertia, namely quintillions and quintillions of bound particles of the epola, "helping" one another to keep the frequency of their vibrations. In inertial effects, the vibrational frequency of the bound epola particles within the accompanying waves of the moving free nuclear particles, or of the constituent particles of the moving atomic body, is proportional to the velocity of the motion. Changing this velocity means also a proportional change of the frequency of the accompanying waves. Any factor, aiming to change the velocity of motion, is confronted not only with the moving particles but also with the quintillions of bound epola particles, tending to keep their vibrational frequency invariant.

The frequency invariance of waves is one of the hundreds of unbeatable proofs that, whatever the medium in which the wave propagates, the medium cannot be "empty", and must consist of discrete particles of matter. Remember that no mathematical equation, no sophisticated wording, no prayers, no donations, not even excessive governmental funding can keep frequencies invariant, only the quintillions of quintillions of vibrating particles can.

This is so for acoustic or sound waves in atomic media, where the vibrating particles in the waves are the atoms, ions, or molecules of the media, and of bodies "immersed" in the media. This is equally so for all kinds of electromagnetic waves, representing bulk deformation waves in the epola. Here the vibrating particles are mainly the bound electrons and positrons of the epola, also of the epola within atomic bodies.

In atomic bodies, low frequency vibrations of the bound epola particles, as in radio waves, are joined by vibrations of the 'free' electrons (especially in metals). At frequencies of light (infrared, visible, and ultraviolet), the vibrations of the bound epola particles are joined by corresponding vibrations of the outermost orbital electrons of the atoms and ions of the atomic bodies. At higher frequencies, electrons on inner orbits become involved, as well as the nuclei of the atoms, which are "immersed" in the epola.

As is the case with all laws and principles, the principle of frequency invariance does not hold "forever". Sufficiently strong factors may be able to change the "attacked" frequency, or even destroy the wave motion. Such are tornado or hurricane winds for sounds, and "magnetic storms" for radio waves. Factors able to cause frequency changes are, e.g.,
    motion of the wave emitter in the wave-carrier (Doppler effect),
    absorption with following re-emission of the waves, and
    the passage of waves through regions of strong EMG fields.
The factors depend on the conditions in the wave-carrying medium.

18.03. The Doppler Effect

In 1842, C.J.Doppler derived, based on the ether wave theory of light, that when the source (emitter) and the receiver (observer) of light move <u>toward</u> each other, then the frequency f_r of the received light is higher than the frequency f, received when they are at rest. This means that the light, with all its spectral lines, is shifted toward the violet or blue color, or to the "blue" edge of the spectrum. Hence, the received light is BLUE-SHIFTED, by a value $f_r - f$. *Vice versa*, when the emitter and receiver move <u>away</u> from one another, then the frequency of the received light is lower than the frequency received when they are at rest. Hence, the received light is RED-SHIFTED by the value $f_r - f$.

The change in frequency, $f_r - f$, divided by the frequency at rest, i.e., the ratio

$$(f_r - f)/f,$$

yields the value of the BLUESHIFT, or REDSHIFT, accordingly. Doppler found that these values are equal to the velocity v_r of the relative motion of the emitter and receiver, divided by the velocity of light c, i.e., they are equal to the ratio v_r/c. The formula for the Doppler effect is thus,

$$(f_r - f)/f = v_r/c.$$

For example, moving in a space vehicle toward the Sun with a velocity of 30 km/s, which is a ten-thousandth of the velocity of light ($v_r/c=0.0001$), we would observe in our spectrometer, that the frequencies of all spectral lines in the sunlight spectrum are blueshifted by a ten-thousandth of their "normal" values. Moving with the same velocity away from the Sun, we would observe a redshift of all spectral lines by a ten-thousandth of their "normal" frequency f.

Doppler's findings were bitterly opposed by the scientific community for two decades, though they could be observed in the increased frequency, or higher pitch of the sound of the whistle of an approaching locomotive, and in the reduced frequency, or lower pitch of the sound of the same whistle, when the locomotive moves away.

The first to accept Doppler's derivation was Ernst Mach, who in 1860 suggested that the Doppler effect might be observed in the spectra of stars. Such observations of the Doppler effect were then carried out in 1868 by W.Huggins. From the values of the red- and blue-shifts in the spectra of stars, their velocities were found to have values starting from 0.3 km/s, and up to 100 km/s. In 1905, J.Stark measured the Doppler shifts in the spectra of canal rays, in which the positive ions moved with velocities up to 1000 km/s, much faster than celestial bodies. At the other extreme, Q.Majorana was able to measure in 1918 the Doppler effect caused by light sources, which were mechanically propelled to velocities of 200 m/s (720 km/h, or 450 mph).

Since World War Two, the Doppler effect is widely used to determine the velocities of various vehicles. This is done by emitting upon them a radar (radio) beam, and measuring the frequency shift in the beam, which is reflected from the moving vehicle, and received back by the radar. With the radar beam, velocities as small as 10 m/s can be measured, with an accuracy of a meter per second.

In water, radio waves are strongly absorbed, and the radar cannot be used. The detection of submarines is done with *sonars*, which emit ultrasound signals and receive the beams, reflected from bodies in the water. Similar ultrasound detection is used by marine animals, and by bats. From the time difference between the start of emission and the start of reception of the reflected signal, and knowing the velocity of sound in the water at the given conditions (say, 1400 m/s), the distance to the reflecting object can be determined. When the object and the sonar move sufficiently fast, away from, or towards one another, then their relative velocity can be determined from the Doppler shift in the frequency of the reflected signal.

18.04. Invariance of the Velocities of Light and Sound in the Doppler Effect

The derivation of the formulae for the Doppler effect, also in textbooks, involves the algebraic (also geometric) addition of the velocities of the emitter and receiver to the wave velocity in the medium, i.e., to the velocity of sound in atomic media or bodies, or to the velocity of light in space (or in atomic fluids or solids, in which the motion occurs). However it is well known that the wave velocity cannot be physically changed due to the motion of the source and/or receiver of the wave.

The Doppler derivation simulates the standard procedure of VELOCITY ADDITION, when *one and the same body* simultaneously participates in several motions, and physically HAS, or moves with the velocities of these motions. For example, the velocity of a sailboat is the sum of the velocities, submitted to the boat simultaneously by the engine, by the current(s) in the water, and by the winds and flows in the air.

We should be aware of the fact, that in the case of the motion of the receiver and emitter away from or toward one another, *there is no physical addition of velocities whatsoever*. The sound is propagating in the air or other atomic bodies with ITS velocity, which depends solely on the physical conditions in the media, and does not depend on whether the receiver or emitter is moving in the medium or not.

Similarly, electromagnetic waves of all kinds are propagating in space, or within atomic media, with the velocity of light, which depends solely on the physical conditions of the epola in space, or within the atomic media, and does not depend on the motion of the emitters and receivers in them.

There cannot be any physical addition of the velocity of sound or light to the velocity of the emitter or receiver, also because the velocity of sound or light is not a velocity of a physical body but of a transfer process of the vibrational or wave energy from one particle of the medium to the next in row.

The emitter, receiver, or any other body, moving in a wave-carrying ambient, may affect the velocity of the waves only in the narrow "channel" surrounding the body. In this channel, actually in very thin layers of the ambient adjacent to the body, the motion of the body may be able to change the physical conditions in the ambient. Thus,

the wave velocity in ambients, apart from the thin layers adjacent to moving bodies, is not influenced by the motion of the bodies.

As an example, consider a locomotive, emitting a sound signal, while moving with a velocity of 33 m/s (about 120 km/hour, or 75 mph), which is a tenth of the velocity of sound. At such, and at even larger velocities, the compressed air layer in front of the locomotive, the rarefied air layer behind, and the layers adhered to the walls and moving with the velocity of the locomotive, are a few inches thick. Hence, several inches away from the locomotive, its motion does not change the velocity of sound, not in the forward, backward, sideward, or any other direction.

Some air currents caused by the motion can be felt in the vicinity, say, at a distance of ten meters away from the vehicle. However, the velocities of these currents are many times smaller than the velocity of the locomotive, and completely insignificant compared with the velocity of sound. Such currents cannot and do not have any effect on the velocity of sound.

18.05. Physics of Frequency Changes in the Doppler Effect and Legality of their Calculation by Velocity Addition

In the Doppler effect, the velocities of light and sound remain invariant, and nothing is added to them or subtracted from them. Nevertheless, the use of the rules of velocity addition to calculate the frequency changes in the Doppler effect, as well as in any other calculation, is perfectly legal. In general,

a mathematical treatment, worked out for a particular process,
may be legally and successfully applied
to the treatment of distinct and even opposed processes,
with gain to both physics and mathematics.

Trouble to physics starts when the **interpretation** of such treatments masks the physical distinctiveness and particulars. One then ends up with the calculation alone, and with no physics left. In the teaching of the Doppler effect, this occurs when the physical reason for the observed and calculated changes in frequency is given as **due** to the velocity addition.

The motion of the wave-emitting body physically increases the number of vibrations per second in the medium (wave-field) in front of the moving emitter, and reduces the number of vibrations per second in the medium (the wave-field) behind the emitter. Hence the wave frequency and photon energy is increased in front of the moving body, and is reduced behind the body. In directions emerging from the body perpendicular to the direction of its motion, the emitted wave frequency is unchanged (i.e., equal to the frequency emitted when the body rests in the medium).

Contrarily, the motion of the receiver (detector) does not change the wave frequencies in the ambient. While moving toward the source, the detector just crosses or counts more waves per second than is the number of waves per second (frequency) formed by the emitter in the field. Moving away from the source, the detector crosses or counts less waves per second than the number of waves per second formed by the emitter in the field.

A more detailed physical analysis of the Doppler effect based on the epola model was given elsewhere and we see no point in repeating it here. It is interesting to note that it leads to the same calculative results as the physically absent addition of velocities.

It is quite often so that models opposed to one another lead to the same mathematical results. Hence,
>   if the mathematics "works", then this does not necessarily prove
>   that the model employed in the derivation is the right one.
Accepting the fact that the derived formula "works"
>   as proof that the employed model is the right one
>   may promote the wrong and "kill" the right.
Or at least bury the quest for the right.

18.06. Frequency Changes in Waves Due to Absorption, Re-Emission, and Reflection

When a wave reaches a boundary with a different medium, then part of the wave energy of the incident beam is reflected back, part is refracting into the other medium, and part is absorbed in the boundary region. By the principle of frequency invariance, one expects the frequencies of the reflected and refracted beams to be the same as the frequency of the incident wave. However, it is not always so.

Waves of certain frequencies, reaching the particles of the other medium, may be fully absorbed by these particles, just because the frequency corresponds to one of the frequencies of the *natural* oscillations of the particles, to one of their "*normal modes*". In the absorbable waves, the energy, transferred from particle to particle of the medium, or the *quantum energy*, corresponds to the energy of one of the "normal modes" (*"eigenvalues"*) of the energy of the absorbing particle.

Particles which absorbed the energy, are during a certain time in a higher energy state, in an *excited state*. The particles may then re-emit the absorbed energy in full, or in part only. If the re-emitted energy is equal to the energy originally absorbed, then the frequency of the re-emitted waves is the same as the frequency of the incident waves, and the *frequency invariance* holds. When the frequency of the re-emitted wave is lower, the frequency invariance does not hold. The pitch of the re-emitted sounds is then lower, and the spectrum of the observed re-emitted light is redshifted.

Similar "redshifting" may occur also in light reflected from colored surfaces, or in light transmitted through transparent colored bodies.

For example, if the surface of a body is illuminated by white light, but appears yellow, then this means that the surface reflects only the yellow light, and absorbs or transmits all the other constituents of the white light. To check if there is any redshifting in this <u>selective reflection</u>, we shine yellow light upon the surface. If the reflected light is also yellow, then the case is of "normal" selective reflection.

If not (i.e., if the surface is dark under yellow light), then we illuminate the surface with green light, of higher frequency and quantum energy. If the reflected light is yellow, then we have a case of redshifting on reflection. Apparently then, the atoms on the surface absorb the green light, undergo a transition into an excited state, from which they re-emit the quanta of yellow light, which are lower in energy.

With increasing distance from the emitter, the amplitude of the vibrations of the particles of the medium decreases. The farther the particle from the emitter, the smaller its amplitude. The vibrational energy of the particles of the medium is proportional to the square of the amplitude. Hence, the vibrational energy of these particles decreases with the distance from the emitter. This is so, *firstly*, because with increasing distance, the wave energy spreads among increasing amounts of particles. The number of these particles at each given distance from the emitting center is proportional to the square of the distance. Hence, particles of the medium, located at a larger distance from the emitter, obtain a smaller share of the emitter's energy.

Secondly, as the distance crossed by the wave increases, the number of absorption events increases, too, and the energy of the wave decreases. Hence, the vibrational energy and amplitude of particles, located farther from the emitter, is smaller also because of absorption, the smaller the larger the distance, and the more there are absorbing factors on the way of the wave. The absorbing factors are also, in addition to the host particles of the medium, dust, faults in the structure of the medium, foreign atoms, molecules, and other "impurities".

18.07. Linear and Non-Linear Absorption

We should distinguish between *linear* and *non-linear* absorption. If the absorption reduces only the vibrational amplitude and energy of the particles in the medium, but the frequency of the wave, hence the quantum energy (energy transferred by single particles) remains unchanged (no redshift or lower pitch), then the absorption is linear. The overall decrease in wave energy with distance from the emitter is then expressed in the reduction of the intensity of light or sound, not in the reduction of frequency; thus in the reduction of the <u>number</u> of quanta (photons, phonons), and not in the quantum energy.

In *non-linear absorption*, the frequency of the wave decreases, too, not only the intensity. Absorption with re-emission of the absorbed energy quanta in full does not directly result in non-linear absorption, but re-emission of quanta of lower energy does. Similarly, absorption with a following reflection of lower frequency waves is a case of non-linear absorption. Generally, waves crossing very long distances undergo very many events of absorption not only on the host particles of the medium but also on structure faults (pockets of increased and reduced density, "holes", etc.), foreign particles, dust, as well as numerous reflections, especially from dust. Some of these processes may be connected with a reduction in frequency.

Even if the number of events of non-linear absorption is insignificant compared with that of linear absorption events, waves received after passing extremely long distances have not only largely reduced intensities, but always have some slightly reduced frequencies. Sounds from far away, received with the help of gigantic microphones and amplifiers, have a slightly lower pitch. Long radio waves, which traveled over our globe, fade to a lower frequency, and light from distant nebulae, detectable with large telescopes, is redshifted.

18.08. Wave- and Propagation Velocities of Absorbable Waves

Another result of the absorption with re-emission is the reduction in the propagation velocity of the waves, compared to the velocity of waves which are not absorbed by the particles of the media. This can be explained by considering the time, during which the wave is absorbed by the particles of the medium, and during which there is no energy transfer, thus no wave motion. When the wave is re-emitted, it again propagates with the WAVE VELOCITY, but the OVERALL *propagation velocity* is significantly reduced because during part of the time the wave energy is not transferred, and the wave motion is delayed.

As an illustration, consider car travel, with the speedometer showing 55 to 60 mph. Hence, you expect to pass 50 miles in an hour or less. However, the actual travel time turned out to be two hours, though there were no disasters on the way. You were simply stopped or "absorbed" at traffic lights and crossings, each time for a while, then "re-emitted" and moving till the next "absorption" event. In car travel, one has to distinguish the velocity of motion, when he really moves, from the OVERALL velocity, found as the quotient of the traveled distance by the travel time. In our example, the overall velocity is 50 miles divided by 2 hours, or 25 mph, half the velocity of motion.

Let us assume, e.g., that a wave which is not absorbed by the particles of the medium, propagates along with a wave which pretty soon is to be absorbed. Then it seems to us that the velocities of the two waves have no physical reason to differ from one another, or to significantly differ from the wave velocity in a non-absorbing medium. But at a certain moment, one of the waves is absorbed, thus out of the velocity competition with the other wave, for a time during which its energy is "hosted" by absorbing particles ("stopped at a traffic light"). The longer this "hosting" time the longer the overall propagation time, thus the lower the overall propagation velocity.

We can verify our presentation by comparing the wave velocities and the effective propagation velocities of light in various media. In clean gases at normal conditions, the velocity of light differs from c by a few hundredths of a percent (~0.05%). Our explanation is that the distances between the molecules in gases at normal conditions are ten to fifteen times larger than the molecular sizes. Hence the "free path" of a wave between consecutive absorption events (as between consecutive red lights in your car travel) can be quite long. Thus the hosting times on absorbing molecules, far away from one another, do not significantly increase the propagation times, do not reduce the velocity of light, and do not diversify the velocities of waves of different frequencies.

Contrarily, in liquids and solids, the atoms, ions, and molecules touch and overlap. The free path of waves between consecutive absorption events is significantly shorter than in gases. Hence, the hosting time of wave energies on absorbing particles may significantly lengthen the propagation time, and reduce the overall propagation velocity of light in those liquids and solids, which are at all transparent to light.

18.09. Propagation Velocities of Light in Atomic Bodies

The propagation velocity of visible light in atomic bodies is:
in normal air - by 0.03% lower than in epola space, thus close to c;
in water - by 25% lower, equal to c/1.33, or 225,000 km/s,
in light glass - by 33% lower, equal to c/1.5, or 200,000 km/s,
in heavy (crystal) glass, by 43% lower, equals c/1.7, or 170,000 km/s,
and in diamonds - by 60% lower, equal to c/2.4, or 120,000 km/s.

The carrier of electromagnetic waves in solids and liquids is still the epola, and not the atoms alone of these bodies. Within these bodies, the epola is slightly deformed by the presence of the electrons and nuclei of the atoms of the bodies. Therefore, the wave velocity of light should be smaller than c, or than the velocity of light in gases, but by less than a percent, if it were not for the hosting times of the wave energy on the absorbing and re-emitting atoms of the bodies.

For example, the density of water is ten billion times lower than of the epola alone. "Filling" a certain volume of the epola space with water constitutes in this volume a density increase by a 10^{-10} part. The velocity of bulk deformation waves is inversely proportional to the square root of the density. Therefore, the velocity of light in an epola volume "filled" with water should decrease by a 10^{-5} part, or by a thousandth of a percent.

A further decrease should occur due to the distortions of the epola units, hosting the nuclei and electrons of the water molecules. In and around these epola units, the binding energy between the epola particles is reduced, also causing a reduction in the wave velocity of light. Altogether, we may expect a velocity reduction of still less than a percent. Therefore, the observed 25% reduction of the velocity of light in water should be ascribed to absorption and re-emission of light in the atoms and molecules.

Indeed, the velocity of X-rays and of gamma-rays, which are not absorbed by atoms, and not hosted by them, is in all atomic bodies, solids, liquids and gases, just very close to c, with no dispersion, i.e., the same for all frequencies of these rays.

18.10. Dispersion of Absorbable Waves

Because the hosting times of quanta of different frequencies is different, their overall propagation velocities in the medium are different, too. This is the main cause of the observed dispersion of sound and light. In our presentation, dispersion is not due to the differing <u>wave</u> velocities, or to a dependence of the wave velocity on the wave frequency, for which there is seldom a net physical reason. Dispersion is due to the different <u>overall</u> propagation times, or <u>overall</u> **propagation** velocities of waves of different frequencies.

We can verify our assumption by considering, that in a clean normal gas, the propagation velocity of all electromagnetic waves is almost the same, i.e., there is almost no dispersion, and that this "one for all" velocity is just insignificantly smaller than the velocity of light c in the "empty" space (in the "clean" epola). There is also very little dispersion of <u>sound</u> waves of different frequencies in clean gases.

The explanation is, as in Section 8, that in gases, the distances between consecutive absorption and re-emission events are large. Even if the hosting times of light quanta of different frequencies do differ significantly, the hosting events are so scarce, that altogether their effect on the propagation of light and sound is negligible.

Similarly, X- and gamma-rays, which are not absorbed by atoms and molecules, not only propagate in atomic bodies with velocities extremely close to c, but also show no dispersion, in spite of the fact that their frequencies differ from one another by hundreds and thousands of times.

Contrarily, the frequencies of visible light differ by a factor of no more than two. However in solids and liquids, the atomic and molecular absorption with re-emission of the frequencies of visible light is so strong, that the hosting times significantly reduce the propagation velocities of light, by 25% in water, and up to 60% in diamond. Then, because the hosting times depend on the frequencies of the light quanta, dispersion is very strong and easily observable.

For example, the rainbow is due to the dispersion of light in small droplets of water, suspended ("hanging") in the air. Stronger dispersion is observed in glass, especially in the heavy "lead" or crystal kinds of glass. In diamonds, dispersion is so strong and (in properly cut ones) so beautiful, that they became "girls' best friends".

The significantly increased numbers of absorption and re-emission events of sound waves in liquids and solids, result in observable dispersion phenomena of sounds in these media. The propagation of sound in liquids, and especially in solids, is also influenced by the order, arrangements, turnings and squeezing of the atoms, ions, and molecules, as well as by the form of the body, by its surface structure and conditions, under which bulk waves of different frequencies are reflected back into the body. Reflections from the surfaces, especially selective reflections of differing frequencies, and the forms of containers (e.g., of theater and concert halls), are important also for the propagation and the qualities of sounds in the air. Then come the large differences (up to a factor of three) in the propagation velocities of sound in different directions, especially in crystals.

All these effects do not exist in the "clean" epola, so that the physics of light propagation in the epola is much simpler. But also when light propagates in atomic solids, i.e., in epola regions deformed by the electrons and nuclei of the atoms of the solid, the structure of the body has a much smaller effect on the light than on the sound. In the propagation of sounds, the atomic body itself is the responsible carrier, and the epola plays a background role only. In general thus, the propagation of sounds in solids and liquids is much more complicated than the propagation of light.

Chapter 19

ASTROPHYSICAL ASPECTS
OF THE EPOLA STRUCTURE OF SPACE

19.01. The Gravitational or Einstein Redshift

In a 1911 paper *"On the influence of gravitation on the propagation of light"*, Einstein referred to his 1905 statement that any energy has mass (by itself, as if energy were a self-dependent material body!), and that if the energy is E, then it has a mass E/c^2. Therefore, if light is propagating near a massive star, then it must be attracted to the star; gravitationally, of course, not the way light is attracted to a massive movie star.

The first result is therefore the BENDING OF LIGHT toward the star. The second result is that the velocity of light moving toward the star must be increasing, due to the gravitational acceleration. By the same token, the velocity of light, moving out or away from the star, also of light emitted by the star, or any massive source, must be reduced due to the gravitational attraction and deceleration.

To an observer of the star, the reduction in the velocity of the light, emitted by the star, can be equally interpreted by a backward motion of the star. But the backward motion of the source should cause a Doppler shift of the spectral lines toward the red edge of the spectrum. Hence Einstein predicted that in the spectra of light from massive stars, there should be a redshift, proportional to the mass of the star. This is the *gravitational* or Einstein redshift.

Einstein calculated the expected gravitational redshift in the spectrum of the Sun to be $2.12 \cdot 10^{-6}$, or two millionths. This means that the frequency of each line in the observed spectrum of the Sun, is shifted toward the red edge of the spectrum by two millionths of the normal frequency value. The gravitational redshift in the spectrum of the Sun is equivalent to the Doppler shift in the spectrum of a light source, which is moving away from us with a velocity of 600 m/s.

Einstein's general formula yields for the gravitational redshift in the spectrum of any star the value

$$2.12 \cdot 10^{-6} \cdot M/R.$$

Here M is the mass of the star divided by the mass of the Sun, and R is the radius of the star divided by the radius of the Sun. Redshifts in the spectra of some stars agree with this formula, but some do not. For "white dwarf" stars, for which large values of the Einstein redshift would be expected, the measurements yield only small values, at best. Though Einstein's formulae for the gravitational redshift and the bending of light do not always yield the right values, both these phenomena are well established physical facts.

19.02. Inadequateness of Relativistic Explanations of Gravitational Redshifts and Bending of Light

The prediction by Einstein of the gravitational redshift and of the bending of light, was made under false pretenses and suppositions. *First*, energies do not have any "self-existence", they cannot have any mass, and do not gravitate. The concept of energy was introduced in physics as a feature of physical bodies, describing qualitatively and quantitatively their ability to perform physical work. Hence, only physical bodies have or do not have energy. Any other use of the term energy is against the physical meaning of it.

Second, light is not a body but a wave motion in the electromagnetic field. As such, it can be diverted from its direction of propagation in deformed regions of the field. Its velocity can also be changed in regions of increased or reduced ratios of energy density to mass density in the carrier. These facts unequivocally witness the physical or epola structure of space, and absolutely reject the emptiness of space. *Third*, motions and velocities of atomic bodies in the electro-magneto-gravitational field, or in any carrier of a wave motion, do not physically add to the propagation velocity of light, or of any wave motion.

In a 1916 paper, Einstein renounced his 1911 explanation of the gravitational redshift, denied the possibility of any change in the velocity of light due to gravitational attraction, and returned to his 1905 proclamation of the velocity of light in space as of a constant.

He also expanded this constancy, unchangeability, and unsurpassability of the "vacuum" light velocity c upon the whole universe.

Einstein's newer explanations of the bending of light and of the gravitational redshift were based on the proclamation that the empty space is deformable, and that the deformation of the empty space near a massive star is causing these effects. In addition, he assumed that photons of the emitted light are attracted to the star, however this attraction does not reduce their velocity (photons must move with the velocity of light!) but reduces their energy. Hence, their frequency is reduced, resulting in the observed gravitational redshift.

Obviously, these are not physical explanations, but an "everything goes" make-believe of an explanation. For one cannot seriously talk about a deformation or deformability of space which is absolutely empty, or assume that anything can be explained on the basis of such an unexplainable make-believe "feature" of the emptiness. On the other hand, the bending of light and the gravitational redshift are really due to the gravitational deformation of the epola, thus of the electro-magneto-gravitational field near massive stars. Hence, a real deformation in space excludes the possibility of its emptiness.

The reduction in the energy of photons due to their gravitational attraction to the massive star cannot be considered as a physical explanation, because the photon is not a particle but an energy amount, transferred in the wave process. Therefore, the photon cannot be attracted to the star, just as the wave cannot. Its path can be bent due to the gravitational deformation, but of the epola, not of an empty space.

Moreover, in an empty space, the "empty-space" photon cannot dispose of its energy, not also of parts of it. But in a strongly deformed region of the epola, there are changes in the distances between the bound particles, thus also changes in the densities of energy and mass. Hence it is possible, that the amounts of energy transferred among the particles may change. This would lead to a reduction in the wave frequency, thus in the photon energy. Therefore, some of Einstein's explanations could be valid, but in the epola, not in empty space.

Because the velocity of light is equal to the square root of the binding energy density, divided by the mass density, and because this ratio may change in the deformed epola, then it is also possible that the velocity of light in the vicinity of massive stars may be reduced. Hence, also Einstein's 1911 original explanation may be valid, but only in the epola, not in an empty space.

19.03. The Mirage Model for the Gravitational Bending of Light

The bending of light and the gravitational redshift, predicted by Einstein, but physically unexplainable by the theory of relativity, are physically explained by the epola model of space. Based on the epola model, the formulae for these phenomena can be derived with simple algebra, as were our derivations for Einstein's mass-velocity and mass-energy relations.

Light beams, moving along a massive star at some distance from it, are passing the distorted epola layers under changing angles; first, on approach, through layers of increasing distortion, then, on recess, through layers of gradually decreasing distortion. By the laws of simple optics, these beams should be bent toward the star.

The physics of the bending of light in the vicinity of massive stars is similar to the physics of the bending of light in the phenomenon of mirage, in which light from the object, seen in the mirage, has to cross on its way to the observer irregularly heated layers of air. With the mirage model, for which the mathematics is well worked out, the complexity of the mathematical description of the bending of starlight should be greatly reduced.

19.04. The Three Tests of Relativity

The bending of light, the gravitational redshift, and the perihelion motion of the orbit of Mercury were devised by Einstein as the Three Tests of Relativity, suitable for experimental examination. These tests did yield results close to the theoretically predicted ones, but only in certain cases. Other measurements and observations yield results, which are quite remote from the predicted values. Thus, the test of the relativity theory cannot be considered sufficient, contrary to what so many books and texts expound. We may quote the 1950 evaluation by Max von Laue, that the relativity theory, though a work of a genius, is not completely proven.

All these test phenomena, predicted by Einstein, were not and cannot be physically explained by the theory of relativity. They are physically explained by the epola model of space. Based on the epola model, the formulae for these phenomena can be derived with simple algebra, as are our derivations for Einstein's mass-velocity and mass-energy relations.

The three *"test phenomena"* of relativity, and several other experimentally substantiated phenomena, worked out mathematically ("theoretically") by relativity and quantum theories, and considered as proofs of these theories, are actually physical proofs of the existence of the epola. It is quite common in science, as pointed out in Laue's "History of Physics", that experimental results are interpreted to fit the ruling theory. When the theory is replaced by an opposing theory, it may occur that facts which were considered as proofs of the previous theory turn out to be equally strong proofs of the opposing theory.

19.05. The Orbit Adjustment Redshift

When an atom moves in the epola, an accompanying wave, corresponding to this motion of the atom, is added to the accompanying wave of each orbital electron, and a wave is created around the nucleus. Hence, the electron orbits in a moving atom are adjusted to its motion in the epola, and differ from the orbits of the resting atom.

For example, all atoms of and on Earth have orbits adjusted to the motions, relative to the epola, of the particular spot of earth to which they belong, as well as to the instantaneous velocities of their thermal and other motions relative to the epola. Considering, however, that the orbital velocities of the electrons on the outer orbits of atoms are 2000 kilometers per second, and that on inner orbits the velocities are much higher, one concludes that most of the "regular" motions of atoms have little effect on their orbits.

In atoms or ions, moving in the epola with velocities of hundreds or thousands of kilometers per second, the adjustment of electron orbits to the motion may cause changes in the transition energies of electrons between the orbits. Therefore, the quantum energies of light, emitted or absorbed in such electron transitions, may differ from the energies, emitted or absorbed in similar transitions, by atoms at rest. Hence, the spectral lines in the spectra of fast moving atoms or ions should be shifted toward the red or blue edges of the spectra.

Without analyzing each particular transition in each atom or ion under its special conditions, we may say that the general trend in physical processes is toward the reduction of involved energies, whenever possible. Thus, the changes in electron transition energies between orbits, due to the motion of atoms, would most probably reduce the transition energy, thus resulting in redshifts rather than blueshifts. We therefore expect that

in the spectra of atoms and ions, moving fast in the epola,
there should be shifts of spectral lines, mostly
toward the lower frequency (red) ends of the spectra.

This frequency shift depends on the velocity of the atoms or ions relative to the epola, and not on whether the emitting atoms move away from or toward us. Hence, it is not a Doppler shift. Nor is it a gravitational shift. We shall refer to it as to the orbit adjustment frequency shift, which should mostly be a redshift, and apparently seldom a blueshift.

19.06. The 3K Blackbody Radiation and the Epola Temperature

In the last decades, the search in astrophysics for low energy radiation is carried out with powerful microwave telescopes, and with radiation detectors kept at millikelvin temperatures (thousandths of a degree above the absolute zero temperature). It was found that we are receiving from all directions in the sky a mysterious microwave radiation, referred to as THE BACKGROUND radiation. The spectral distribution of this radiation corresponds to the radiation of a blackbody, the temperature of which is about 3 degrees above the absolute zero, or 3 kelvin (3 K). More recent measurements yielded 2.75 K for the temperature of the mysterious blackbody, emitting this radiation.

The theoreticians of the Big Bang creation theory of the universe worked out several speculations to explain the origin of the 3K blackbody radiation. The most popular of them, and the closest one to physics, so that a natural physicist can at least relate to it, is that the 3K radiation was left over by the Big Bang explosion, and distributed in the empty space all over the universe. This reminds baron Munchausen's story about a hunt at such a bitter frost that the sounds of the trumpets froze. Later when the sun "defrosted" the left over sounds, the empty air in the forest was trumpeting.

In natural physics, microwave radiation (or sound) cannot just "stay". It must propagate, and for this it needs a physical carrier and emitter. If the emitter was the spot where the Big Bang started, then all kinds of its radiation would have passed our region quite a few quadrillennia ago, with one and the same velocity of all (there is no dispersion in the empty space!). If the radiation comes back from reflecting borders in the universe, then what and where are these mirrors, and why do they reflect the 3K microwaves only?

In natural physics, it is clear that the 3K radiation can only be emitted by a body, the temperature of which is 3 kelvin. Because there are no atomic bodies besieging us from everywhere, we must consider the epola itself as the source of this radiation. Hence, we say that

the 3K microwave radiation, reaching us from everywhere in the skies, is the thermal radiation of the bound electrons and positrons of the surrounding epola, and is due to the random (thermal) vibrations of these bound particles around their lattice sites.

Being the radiation of the surrounding epola, the 3K radiation is not a background radiation in the universe, but a foreground radiation of the epola space in front of us. Hence, the temperature of the surrounding epola is around 3 kelvin. It is clear to a natural physicist, that whatever we find in our region of the epola, or in our backyard, is not necessarily universal. Hence, there must be epola regions of other temperatures. For example, in regions of more intense nuclear activity, the epola temperature is higher, and *vice versa.*

19.07. The Field Temperature Redshift

Let us consider that there is a "neighboring" epola region, in which the absolute temperature of the epola, or of the electro-magneto-gravitational (EMG) field is 30K, tenfold higher than in "our" 3K epola. Let there also be another neighboring region, in which the field temperature is tenfold lower, or 0.3 kelvin.

Then it is important to note that there are no means by which we can receive the microwave blackbody radiation of the neighboring fields. According to the laws of spectroscopy, the blackbody radiation of neighboring epola regions is completely absorbed on the borders of our epola. The "background" radiation which we can receive is solely the 3K blackbody radiation of the epola in front of us. We <u>can</u> receive the radiation from atomic or nuclear bodies (sources) located in other regions, only if the radiation of these bodies corresponds to temperatures higher than the temperature of "their" surrounding epola, and higher than the temperatures of all epola regions (EMG fields) on the way to us.

Let us now assume that in each of the regions, there is an atomic source of radiation, a blackbody, the absolute temperature of which is 1000 kelvin. If all these radiations reach a region in which the epola temperature is closest to zero kelvin, then the radiation of the 1000 K source from the 30K epola temperature region, appears there as radiation from a source at 970 K. The radiation of the 1000 kelvin source from our epola region, appears there as the radiation of an atomic source, kept at a temperature of 997 kelvin. Finally, the radiation from the 1000 K source, located in the 0.3 K region, appears in the close to zero kelvin epola region as the radiation of an atomic source, kept at a temperature of 999.7 kelvin.

If the radiation from the atomic source in the 0.3 K region is received by a spectroscopist in our region, then he will find that this radiation has a higher energy and frequency than the radiation of the source in our region, i.e., that the received radiation is blueshifted. The spectral analysis of the radiation from the 30K region would show that this radiation has lower energies and frequencies than the radiation of the identical source in our region, i.e., that the radiation, received from the 1000 K source, located in the 30K epola region, is redshifted.

The frequency shifts of atomic radiation reaching us from regions of higher or lower epola temperatures, or of higher and lower temperatures of the electro-magneto-gravitational (EMG) fields, will be referred to as FIELD-TEMPERATURE frequency shifts. We may thus say that atomic or nuclear radiation, reaching us from epola regions, where the EMG field temperature is lower than 3K, is received blueshifted, and radiation from regions of field temperatures higher than 3 K, is received redshifted.

We note that among the field-temperature frequency shifts, redshifts prevail, because for separate epola regions, there are many more options to exist at temperatures above 3 K than below 3 K. This is so, because there are certain minimum steps in the temperature differences of separate regions, below which the regions cannot be distinguished, and are not separate. For example, the step can be, depending on conditions, 0.1 K, 0.01 K, or a millikelvin, etc. But whatever the step, there are always many more steps above 3 K than below.

19.08. Temperature and Impurity Effects in the Cosmic Epola

The epola temperature is elevated in regions of the universe, in which there are more "hot" stars, where there is more nuclear activity, more emitted nuclear particles, and a higher density of electromagnetic radiation of all kinds. The dependence of the velocity of light on the temperature of the epola may be expected to be similar to (but much simpler than) the temperature dependence of the velocity of bulk deformation waves in solid lattices, particularly in the unbounded sodium chloride crystal. As a rule, the velocity of light should be smaller where the epola temperature is higher.

Where the epola temperature is so high that the average (thermal) energy per particle becomes equal to the binding energy, the lattice melts, turning into a liquid, which consists mostly of electron-positron pairs. By laws of thermodynamics, the epola should melt at a temperature of six billion kelvin. At even higher temperatures, the epola may turn into a gaseous mixture of electron-positron pairs and free electrons and positrons. The velocity of light is then drastically reduced, as is the velocity of sound in gases, compared to its velocity in solids or liquids.

In the evaporated epola, the transfer of radiation energy can no longer be described by photons. Just as there are no phonons in gases, there are no photons in the gaseous epola. Therefore, Planck's law, as well as other quantum or epola radiation laws, do not hold. Such may possibly be the situation inside certain kinds of black holes, and in other unexplained forms of matter in the universe.

The propagation of electromagnetic radiation in different regions of the universe should be affected by imperfections and impurities in those regions, and should depend on the kinds, concentrations, and spatial distributions of these factors. Imperfections are distortions or defects in the epola per se. They include unoccupied lattice sites (*vacancies*), bound electrons and positrons positioned out of lattice sites (*interstitials*), dislocated lines or planes of host particles, domains, grains, and boundaries between various formations of the particles in the lattice.

Distortions of the lattice can be caused in the vicinity of large masses of atomic or nuclear bodies. In the distorted layers of the epola, the velocity of light should be reduced, the more the larger the distortion. The distortion of the epola layers is the larger the closer to the massive body. Hence, a light beam moving in a radial direction <u>toward</u> a massive star, propagates with increasingly increasing velocity. Radial beams, moving away from the star, or beams emitted by the star, propagate with a decreased velocity, which increases toward the "normal" value with increasing distance from the star.

Impurities in the epola are nuclear particles other than bound electrons and positrons. Hence, impurities include free electrons and positrons, free electron-positron pairs, protons, neutrons, various atomic nuclei, and thousands of kinds of unstable nuclear particles. Impurity particles in the epola strongly distort the lattice units around themselves, pushing the bound particles far out of their lattice sites.

We should notice that the positron or electron mass is uniquely small among nuclear particles: the mass of the next lightest stable nuclear particle, the proton, is 1840 times larger. Therefore, it is hard to imagine that any of the known stable nuclear particles may replace (substitute) an electron or a positron on its lattice site in the epola, as impurity atoms or ions replace the host atoms or ions in solid lattices.

High concentrations of impurities in the epola can be caused by nuclear particles, ejected into an epola region by cosmic nuclear reactions. Such concentrations may tear the epola apart, creating another kind of black hole, in which quantum or epola radiation laws would not hold.

Chapter 20

COSMOLOGICAL ASPECTS
OF THE EPOLA STRUCTURE OF SPACE

20.01. The Hubble-Humason Redshift and the Big Bang

In his paper "*Cosmological considerations on the general theory of relativity*", which appeared in 1917, Einstein applied relativity to the study of the universe. He created hypotheses concerning the distribution of matter in the universe, the possible finiteness of the universe, its time of existence, its contractions and expansions.

Einstein's ideas about the permanently expanding universe, received great impetus from the 1929 discovery by E.Hubble and M.L.Humason, that the spectra of light from distant nebulae (later found to be galaxies} are significantly redshifted. Knowing the approximate or assumed distances to each observed nebula, they also found that the redshift is the larger the longer the distance to the nebula, hence that
*the observed redshifts in the spectra of distant nebulae
    are roughly proportional*
to the approximate distances from the nebulae.

This is the actual physical law discovered by Hubble and Humason, the law of the PROPORTIONALITY OF THE REDSHIFT TO THE DISTANCE, CROSSED BY THE LIGHT ON ITS WAY FROM THE DISTANT SOURCE.

Under normal condition, an unprejudiced scientist, having discovered that something is proportional to the distance, and trying to find why it is proportional, would have looked to what is happening to that something <u>along</u> the distance. Hence, the unprejudiced Hubble, and the critical Humason, would be looking to what is physically acting on the nebular radiation on its way to us. And they could easily point their fingers toward the two then well-known physical phenomena which can, and do cause the redshift. These are the non-linear absorption, and the gravitational deformation of the EMG field.

Unfortunately, Hubble was prejudiced, Humason was submissive, and they were both pressured by the ideas, ideologists, and administrators of the expanding universe. It was a priori decided that the universe is expanding, therefore the nebulae must be running away from us, and if so, then the redshift is a Doppler redshift of sources of radiation, which move away from the observer. Immediately, as in the vicious circle, this argumentation was (and still is) turned over to say, that the spectra of distant nebulae show a large DOPPLER (!!!) redshift, hence this redshift PROVES (???) that the nebulae run away from us, hence the universe is expanding.

The Hubble-Humason law of the proportionality of the redshift to the distance was therefore twisted to fit the ideology, and was formulated to say that the RUNAWAY VELOCITIES OF NEBULAE ARE PROPORTIONAL TO THE DISTANCE to them. This means that the farther the nebula is from us, the faster the poor thing must run away from us, in the name of the ideology.

To fit the physical law of the proportionality of the redshift to the distance, the imaginary runaway velocity of everything which is at a distance of one **megaparsec** from us, has to be around 100 km/s. The megaparsec (symbol **Mpc**), is ~30 quintillion kilometers, or 210 billion times our distance from the Sun, and is traveled by light (at the speed of 300,000 km/s !) for 3.3 million years. In short,

1 Mpc = $3.3 \cdot 10^6$ "light years" = $2.1 \cdot 10^{11}$ "astronomical units" = $3 \cdot 10^{19}$ km.

Everything which is at a distance of, say, ten megaparsec, must run away from us with a velocity of 1000 km/s; from a distance of a giga-parsec, all runs away from us with a velocity of 100,000 km/s, and so on (and there is lots to go on to). The proportionality coefficient in this "twist the physics" formulation, is the "Hubble constant" H, which is, approximately, a hundred kilometers per second per each megaparsec in distance,

$$H = (100 \text{ km/s})/\text{Mpc} .$$

With the development of astrophysical instrumentation, more distant nebulae were discovered, and all the poor little runaways turned out to be galaxies, i.e., tremendously large collectives or conglomerates of billions of stars, and other cosmic bodies. They all showed the Hubble redshift in their spectra, redshifts roughly proportional to their distance from us. When the "penetration" depth of our telescopes into the universe exceeded 50 gigaparsec, enforcing the Dopplerian interpretation meant that the galaxies must run away from us with velocities exceeding the velocity of light.

Amazingly, this did not frighten the expanders of the universe, and it does not frighten them even now, when the depth exceeded 500 gigaparsec. They progressed from an expanding universe to an exploding one, allegedly created in the Big Bang explosion, from a subpoint in space in an octillionth of an octillionth of a second, as many billions of years ago as the particular calculator likes. The universe is still exploding, and will do so for such and such number of years, just to fit the calculations.

20.02. Non-Universality of the Vacuum Light Velocity

The development of the Big Bang theory is based, in addition to the twisted Hubble-Humason law, on illegal extrapolations of two postulates of relativity. First, that the velocity of light, measured here on a 47 mile distance, MUST be the same in the whole now million megaparsec wide observable part of the universe. Second, that the laws of physics, established in our backyard, MUST be the same and equally valid in the whole universe. Clearly, what is good on Earth and may be, within a distance of several light years, is not necessarily right at distances of a trillion light years away.

Being the velocity of bulk deformation waves in the epola, the vacuum light velocity depends also on local conditions in the epola, analogously to the velocity of sound in solid lattices. Such additional conditions are, mainly, the lattice temperature, and the concentration or distribution of foreign particles, impurities, lattice faults or imperfections, etc. The vacuum light velocity c is certainly a very important constant in our epola region, and as far out as the mentioned conditions remain stable, but not in the whole universe.

Radiation entering our uniform epola region from regions where it had a different velocity, propagates here with velocity c, corresponding to conditions in our region. Figuratively, radiation does not "remember" its previous velocities. It always reaches our apparatuses with the velocity c, which corresponds to the conditions in "our" epola, independent of what the velocity is (or was) in the regions where the light was generated, and which it had to pass before reaching our region.

The first to postulate that the vacuum light velocity is a universal constant was H.Poincaré. He did it in order to adjust to the "no ether" results of the Michelson-Morley experiment, and this postulate was taken over by Einstein. In the epola model of space, there is no need for any adjustment to these results, because they are clear and obvious. With only a 10^{-15}th part of its volume filled by nuclear particles, and with the rest of the volume as empty as space, the moving Earth cannot cause winds in the epola space. Thus the epola model solves all problems of light propagation and motion of bodies, without demanding that physical laws and quantities, established and measured in our backyard, must be the same in the whole universe.

20.03. Gravitational Presentation of the Hubble Redshift

We have discussed several phenomena causing redshifts in radiation spectra, which act along the distance, passed by the radiation. These phenomena could be the real physical causes for the Hubble-Humason redshift, which is nearly proportional to the distance from the remote nebula or galaxy.

The simplest of all factors, which cause a reduction in radiation frequency along the path of the radiation, is the non-linear absorption. However as long as the distribution and concentration of the impurities and imperfections in space along the path of the radiation is not known, we can only cite the factor, but cannot do any calculations. Hence, it is not quoted in cosmology (recall the "bicycle stability syndrome", Section 6.10). We are slightly better off with the gravitational distortion of space, which is also reducing the radiation frequency along the path.

It was derived by Einstein in 1911, that radiation passing through a gravitationally distorted region of space loses energy with a reduction in frequency. The gravitational redshift in the spectrum of the Sun corresponds to 600 meters per second of recessional speed, i.e., to a Doppler redshift which would be obtained _if_ the Sun were moving away (recessing) from us with the velocity of 600 m/s. This redshift is caused by the gravitational distortions along the distance from the Sun, thus along one astronomical unit.

Observed nebular redshifts correspond to recessional speeds of the order of 100 km/s along a distance of one megaparsec, or 210 billion astronomical units. Hence, the corresponding recessional speed along one astronomical unit is 100 km/s divided by 210 billion,

$$(100{,}000 \text{ m/s})/2 \cdot 10^{11} = 5 \cdot 10^{-7} \text{ m/s},$$

or 500 billionths of a meter per second. This recessional speed is a billion times smaller than the recessional speed, corresponding to the gravitational redshift in the spectrum of the Sun.

Hence, gravitational redshifts, equal to the observed Hubble redshifts, can be caused by a density of gravitational distortions per one astronomical unit of distance, even a billion times smaller than the gravitational distortions caused by the Sun. We may therefore say that,

if all celestial objects, located along the path of radiation from a distant nebula or galaxy, would create, per astronomical unit, an average gravitational distortion, which is a billion times smaller than the distortion created by the Sun, then this would already be sufficient to consider the Hubble redshift as a gravitational redshift.

In our discussion, we avoided mentioning the epola as the physical medium in which the gravitational distortion occurs. We did it deliberately, to show that the same explanation could be given in 1929, when the Hubble-Humason redshifts were discovered (eighteen years after Einstein introduced the gravitational redshift), and any time thereafter, also by believers in the empty space. The addition of the non-linear absorption as a redshifting factor could also be done then, and with the proper mathematical treatment, this would have helped to develop a more real physical picture of the world.

20.04. The Non-Constancy of the Hubble Constant

The apparent runaway velocity of nebulae or galaxies is calculated from the observed redshift in the spectrum of the object, using the formula of the Doppler effect, and assuming that the velocity of light is the same in all directions, in all regions and spots, everywhere in the whole universe. The distance to the galaxies is approximated from their sizes, types, and other observations.

After years of such practice, it turned out that in the spectra of galaxies, which are equally distant from us, the redshifts are not necessarily identical. Depending on the direction to the galaxy, on its type, or some other, often unknown reasons, the redshift values may differ by up to a factor of five. Hence the runaway velocities, calculated from the redshift values, may also vary by up to a factor of five.

Allowing different values of the runaway velocity in different directions, but at the same distance, means trouble to the expanders and exploders of the universe. They want the universe to expand uniformly in all directions; this is what mathematics likes, thus they believe that it is so. Therefore, it was democratically decided in the name of the whole universe, that all objects located at the same distance, must run away with the same velocities. However, to fit the Hubble law, the objects are allowed to have different values of the Hubble constant.

The Hubble law, forced upon the Hubble-Humason redshifts, says that the runaway velocity v of everything in the universe, which is now located at a distance l from us, is equal to the Hubble constant H, multiplied by l. The formula for the law is

$$v = H \cdot l .$$

To fit this law, a galaxy at a distance l, the redshift of which is five times larger than the redshift of another galaxy at the same distance, should either have a velocity that is five times higher, and an equal Hubble constant, or it may have the same runaway velocity, but a Hubble "constant", that is five times larger, and is then not a constant any more.

To keep the universe exploding uniformly in all directions, the constancy of the Hubble constant was sacrificed, and its calculated values vary now for different galaxies from 30 km/s per megaparsec, to 150 km/s per megaparsec. Such a non-constant "constant" is at best a "factor" only, so that we should be talking now about the Hubble factor. The published reasons for this non-constancy are mathematical, quite speculative, often contradicting one another, and never having to do with natural physics.

During the last few years, the mathematics of the exploding universe was sufficiently developed to manage non-uniform expansions also. The universe may thus explode not as a spherical balloon, but as special toy-balloons made of soft latex, in which it is possible to blow or twist out tentacles or "sausages" of desirable lengths, thicknesses, and directions. This mathematical development is announced as an almighty problem solver. But it seems that all this is just an expensive childish game. It succeeds in drawing attention and funds, possibly because administrators fear that if they will not provide the requested support, then during their term in office the first runaway galaxy might be caught by the competitors.

20.05. Physical Explanation of the Non-Constancy of the Hubble "Constant"

Let us recall that the actual physical law, discovered in 1929 by Hubble and Humason, is that the redshift values in the spectra of distant nebulae, later found to be galaxies, is proportional to the distance l to them. The formula for this law is

$$r = k_H \cdot l,$$

where r is the redshift value, and k_H is the "physical Hubble coefficient", opposed to the Dopplerian Hubble constant H. This coefficient is a constant, when all objects, which are at the same distance l from us, show the same redshift values in their spectra.

In the Hubble-Humason measurements, and in later ones, until the 1960s, the coefficient k_H was more or less constant. (Also, if a measurement proved to the contrary, then the researcher might be either ashamed or afraid to report it; and if he did, it could have been rejected). Therefore, the apparent runaway velocities, calculated from the redshift values with the Doppler effect formula, yielded a constant value for the Hubble constant H.

The more recent powerful means allowed us to extend farther out the observable borders of the universe. It turned out then, that the redshifts in the spectra of equally distant galaxies may differ by a factor of five, depending on the direction to the galaxy, on its kind, and on other factors, also unknown ones. Then the apparent runaway velocity, which is a calculated second-hand derivative of the redshift value, turned out not to fit the Dopplerian Hubble law. To make the runaway velocity fit the law, the Dopplerian Hubble constant was made variable.

Hence, the reason for the non-constancy of the calculated Hubble constant H, which is a third-hand derivative of the redshift values, is that the redshift values in the spectra of equally distant galaxies are not equal to one another, but differ by up to a factor of five. Hence, the physical problem is WHY these redshift values differ. And the answer is that in addition to the Doppler effect, there are at least four other physical phenomena which act in space and cause redshifts (seldom blueshifts) in the passing radiation waves.

The first phenomenon is the non-linear absorption of light. The frequency and energy reduction due to this absorption is proportional to the distance passed by the radiation, if the absorbing bodies, i.e., cosmic dust, absorbing gases, impurities and imperfections in the epola, etc., are uniformly distributed along the path. The redshift caused by this absorption may depend on direction, if the concentrations and distribution of the absorbing bodies differ with direction.

The second phenomenon is the gravitational distortion of the epola space. The gravitational redshift is proportional to the distance passed by the radiation, if the distribution of massive stars along the path is uniform. The gravitational redshift should have a profound directional dependence because of the not-so-uniform distribution of massive stars in different directions. For example, if the radiation from a galaxy passes in a direction, in which there are more massive stars than in the direction from another, equally distant galaxy, then the component of the gravitational redshift in the redshift of the first galaxy is larger.

The third factor is the orbit adjustment in fast moving light-emitting atoms and ions. The resulting orbit-adjustment redshift (seldom blueshift) depends on the speeds of the light-emitting atoms relative to their surrounding epola, and not on the direction of their motion. This redshift does not depend on the distance from us or on the direction to the source. It depends on the kind of galaxy to which the emitting atoms belong, expressing the character of motions in the particular galaxy, its structure, distribution of masses in it, as well as the conditions of the epola in and around the galaxy and its parts.

The fourth factor is the difference in the epola temperatures in the regions where the radiation was generated, and which it had to pass on its way to our telescopes. The resulting field-temperature redshift (seldom blueshift) may increase with distance, because of the probably increased number of epola regions and temperature differences crossed by the radiation. It may also be depending on direction, if the distribution of epola regions of differing temperatures differs with direction.

The Doppler effect is the only one which may result in blueshifts as well as in redshifts, because the universe has no physical reason to be exploding, and galaxies move "toward us" as massively as they move away. They also never move, relative to the epola in and around them, with velocities exceeding a few hundred kilometers per second, or else they turn into fragments. The Doppler redshifts and blueshifts contribute to the diversity of the observed redshift values, which consist of several frequency shifts (also unlisted and unknown ones). The diversity of the observed combined redshift values leads, in turn, to the non-constancy of the calculated Hubble constant.

20.06. Dismissal of the Runaway Interpretation of the Hubble-Humason Redshift, and of the Big Bang

The interpretation of the Hubble-Humason redshift as due to a runaway of the light sources (nebulae or galaxies) from us does not follow directly from the original observation that the redshift values are proportional to the distance from the sources. One may say that it is a second-hand derivative of the original observation. To introduce this interpretation,

first, one has to believe that the universe is expanding in an absolutely empty space, and he has to be ready to twist physical facts in the name of his beliefs;

second, he has to disregard all physical factors, acting on the radiation along its path, has to choose the Doppler effect, which has no connection with the proportionality of the observed redshift to the path length, and has to pronounce that the nebular redshift is a Doppler redshift;

third, after having convinced everybody and himself that the redshift *is* Dopplerian, he turns around and announces that *the Dopplerian character of the redshift proves that the nebula is on a runaway from us*!

We may ask why people should choose such a circling around the experimental result, such a *"via dolorosa"* of speculations around it, instead of considering effects directly connected with the observation. And the answer is that it fits their interests. Job, promotion, prestige, and financial interests.

We may mention that Humason himself was not too orthodox about the Dopplerian interpretation. In a paper published two years after the 1929 discovery of the nebular redshifts, Humason wrote:

"It is not at all certain that the large redshifts observed in the spectra are to be interpreted as a Doppler effect but, for convenience, they are interpreted in terms of velocity and referred to as apparent velocities."

Humason paid a price for his non-orthodoxy. The discovery, the law, and the redshift, were quoted for decades as Hubble-Humason's, but then the Humason name disappeared. His name and contribution were "forgotten", and everything was and is now ascribed to Hubble. Humason was thus removed from the "Hall of Fame" of science, and his name does not even appear in the "who is who in science".

For us to agree with Humason's quoted saying, we would have to slightly sharpen it, to say that

it is certain that the redshifts observed in the spectra <u>should not</u> be INTERPRETED *as due to the Doppler effect.*

We would also have to replace the second use of the word "interpret" by "represent", to say that,

for convenience, the redshift may be <u>REPRESENTED</u> *mathematically in terms of apparent runaway velocities.*

We too used the Dopplerian runaway presentation for the gravitational redshift in the spectrum of the Sun (Section 19.01). However when one uses a mathematical presentation, corresponding to a different phenomenon, he should take all possible care to prevent the turning of the calculative "convenience" into an accepted interpretation, which then replaces and "kills" the physics.

The reason why there was and is such a strong interest in the Dopplerian interpretation, is that with it, the nebular redshift turns into the only one available "experimental proof" that the universe is expanding, and into the main support of the Big Bang theories of the creation of our ever-exploding universe. These theories are more exciting than star-wars' movies. They open opportunities to people with mathematical skills, access to advanced computers, and ability to disregard and twist reality (as do some "modern" painters), to become "creators of universes", to compile new exploding processes for the universes, to invent new exotic particles bearing their names, or introduce more dimensions to the troubled world.

Contrarily, natural physics, with the epola presentations in it, is very prosaic. It restores, introduces, and enforces strict physical rules, language, and thinking. It eliminates the possibility of star wars, of shrinking and blowing up kids, and of twin brothers and sisters traveling between galaxies. With the disproof of the Dopplerian interpretation of nebular redshifts, and with the reversal of the 3K "background" radiation into the "foreground" radiation of the epola in front of us, natural physics turns the Big Bang theory into an unnecessary fantasy.

20.07. Dismissal of the Creation Mechanism of Stars and Nuclear Matter by Gravitational Collapse

In addition to the Big Bang theory of the creation of the universe, the "accepted scientific knowledge" in cosmology is that material bodies were formed due to the gravitational attraction between the particles of diluted clouds of gases and dusts in some regions of the universe. Under favorable conditions, when the clouds reach certain volumes and densities, the gravitational forces may start a collapse of the cloud, leading to the creation of a star. The gravitational collapse is then not only reducing the volume of the cloud and star but also causes their heating, up to the observed temperatures of stars.

The temperatures of the stars are thousands and tens of thousands of degrees on their surfaces, and millions of degrees in their inner cores. Such temperatures, and the tremendous amounts of energy radiated by a star, could only be achieved in nuclear reactions. Nevertheless, mathematics was able to overcome this difficulty and to "prove" that the heating of the star and the replenishment of the increasing amounts of energy emitted by the star are all due to the gravitational collapse, continuing until the star turns into a *dwarf*, *white* or *red*.

The mechanism of gravitational collapse was invented and introduced long before the discoveries of radioactivity, of particles of nuclear matter, and of nuclear reactions. Therefore the gravitational collapse was the only thinkable process, considered able to provide the energy of the stars, with the support of some chemical reactions and burning of the materials in the star. But this did not seem sufficient for the stars, also for our Sun, to last as long as they supposedly do, and will.

It is therefore amazing that the gravitational collapse is still considered and thought in cosmology as the creator of stars, provider of their radiated energy, and even as a creator of nuclear particles and nuclear stars. The reason seems to be that the gravitational collapse is easy to "explain" and to calculate. Hence, we again face a case of the "bicycle stability syndrome".

We should mention two more difficulties connected with the gravitational collapse. First is that space (even within the Solar System) is extremely empty of atomic matter. Hence, to create a star out of the emptiness, atoms, molecules, and dust particles would have to start collapsing from distances trillions of times larger than the diameters of stars.

The second difficulty is the extreme weakness of the gravitational forces compared with the electromagnetic forces. Let us recall that the electrostatic and magnetic attraction and repulsion between nuclear particles, or between ions, is 40, 30, and 20 orders of magnitude stronger than the gravitational attraction. Therefore, the gravitational collapse may act in coagulating large lumps of atomic matter into atomic bodies, but is hardly effective in bringing molecules together to form a body. Gravitational forces cannot overcome the slightest deviations from electric and magnetic neutrality in atoms and ions.

Therefore, the gravitational collapse cannot create molecules, certainly not atoms, and has no influence whatsoever on the creation of nuclear particles, and not on their behavior. Not even of the electrically neutral ones (neutrons, neutrinos). Here the epola model of space can provide a physical mechanism, which is the epola collapse.

20.08. Epola Collapse and Creation of Atomic Matter, Nuclear Matter, and Black Holes

A strong effect of nuclear action may ignite the epola to collapse locally by forcing the bound electrons and positrons to such small distances, that the short-range repulsive interaction is either reduced or cancelled. Then the particles may approach each other so closely that the nuclear forces start acting and will coagulate them. Such local collapse may create unstable nuclear particles, the smallest of which, the muon, consists of 140 electron masses.

To create a neutron or proton, 1840 bound electrons and positrons would have to collapse. Thus, 1840 epola unit cubes would have to be emptied, creating around the newborn particle an empty hole of a diameter of 65 fermi, 150 times smaller than the diameter of the hydrogen atom. To create, e,g., the nucleus of a copper atom, an epola sphere of diameter 250 fm has to be emptied. This is 80 times smaller than the diameter of the copper atom, and the volume of the empty hole is only two millionths of the volume of the atom. Hence,

the creation of atomic matter by a local epola collapse does not require any draw-in of particles from distant regions, and the regular epola structure in and around the newly created atomic body can be promptly and smoothly restored.

A local collapse in the epola may also initiate the creation of a new celestial body. Due to the high density of matter in the epola, the creation even of large bodies of atomic matter is possible everywhere in the epola "on premises", provided that the needed nuclear igniting action is there. Thus, in order to create a celestial body, there is no need to collect matter from vast volumes of the universe, as it is assumed in existing hypotheses of gravitational collapse.

The creation of an extended <u>nuclear</u> body by an epola collapse would require the emptying of a very large volume of the epola around the nuclear body. The question arises if the epola structure in this large emptied volume can be restored, and if so, then how long it would take. During the apparently long process of restoration, the volume would contain a more or less diluted gaseous mixture of electrons, positrons, and electron-positron pairs.

This more or less empty volume in the epola would not emit light nor would it be able to transfer electromagnetic radiation energy, as does the epola. Hence, the emptied volume would not be transparent for light. Directing a telescope toward this spot we would not be seeing anything behind it, no starlight, no other electromagnetic radiation. Analogously, a volume, out of which the air was pumped out, does not produce sounds and does not transduce them. Obviously, we got the third epola scenario for the creation of a black hole.

The distortion of the adjacent epola around this black hole could be detected as a gravitational distortion, the larger the larger and emptier the hole. The degree of emptiness or blackness of the hole may reveal its age, because with the passing of time, more and more particles are drawn into the hole, and it becomes less empty, less black, and more "gray". The size of the black hole actually reflects the mass of the created nuclear body.

The nuclear body could, in the meantime, leave the hole. But if the body is inside the hole, then it cannot be seen or otherwise detected electromagnetically, because there is no epola around it to carry radiation. Nor can the mass of the body be detected gravitationally, because gravitation is also carried by the epola. Therefore, the information about the mass of the created body can only be obtained from the size of the emptied epola volume (from the size of the black hole), and also from the apparent mass of the black hole, detected via the gravitational distortion of the epola around the black hole. This apparent mass may also contain information on the time assed since the collapse.

AUTHOR INDEX

Ampére, André Marie (1775-1836); 139
Andersen, Hans Christian (1805-1875); 5, 71, 117
Anderson, Carl David (1905-); 1, 86, 87, 187
Aristarchus of Samos (3rd Century BC); 73
Aristotle (384-322 BC); 71, 73

Balmer, Johann Jakob (1825-1898); 193
Bohr, Niels (1885-1962); 85, 194-195
Bragg, Sir William Henry (1862-1942), and
    Sir William Lawrence (son, 1890-1972); 101
Brahe, Ticho (1546-1601); 76
Broglie, see de Broglie

Cavendish, Henry (1731-1810); 125
Compton, Arthur Holly; 162
Copernicus, Nicolas (1473-1543); 75
Coulomb, Charles Auguste de (1736-1806); 105, 134

de Broglie, Louis Victor, Duke (1892-1987); 6, 44, 122
Doppler, Christian Johann (1803-1853); 223

Eddington, Arthur Stanley, Sir (1882-1944); 67
Einstein, Albert (1879-1955); 4, 21, 45, 66-67, 184, 185, 187, 248
Faraday, Michael (1791-1867); 139, 183, 218
Feynman, Richard Philips (1918-1989); 67, 206
Fraunhofer, Joseph von (1787-1826); 178, 179

Galileo Galilei (1564-1642)); 61, 66, 125
Gilbert, William (1540-1603); 138
Goudsmit, Samuel Abraham (1902-1978); 140

Hawking, Steven; 67, 68
Herschel, Friedrich Wilhelm (1738-1822); 177
Hertz, Heinrich (1857-1894); 183, 185
Hubble, Edwin (1889-1953); 245, 251, 254
Huggins, Sir William (1824-1910); 224
Humason, Milton LaSalle (1891-1972); 245, 251, 253-254
Huygens, Christiaan (1629- 1695); 4, 182

Kepler, Johannes (1571-1630); 76-77, 125, 200

Laue, von, Max Theodor Felix (1879-1960); 101, 238, 239
Lavoisier, Antoine Laurent (1743-1794); 20, 23, 176, 210
Lenard, Philipp (1862-1947); 186

Mach, Ernst (1838-1916); 224
Madelung, E.;105
Majorana, Quirino; 224
Maxwell, James Clerk (1831-1879); 4, 66, 139, 182, 218
Michelson, Albert Abraham (1852-1931); 4, 82
Morley, Edward Williams (1838-1923); 4, 82
Munchausen, von, baron; 241

Newton, Sir Isaac (1643-1727); 3,4,19,56,65-66,125,177-178,194.

Oersted, Hans Christian (1777-1851); 138

Pauli, Wolfgang; 146, 204
Planck, Max (1858-1947); 67, 159, 184, 185
Poincaré, Henri (1854-1912); 247
Ptolemy, Claudius (2nd Century AD); 74, 77

Ritter, Johann Wilhelm (1776-1810); 178
Rutherford, Lord Ernest (1871-1937); 2, 6, 33, 46, 84, 189

Stark, Johannes (1874-1957); 224
Uhlenbeck, George Eugene; 140
Wollaston, William Hyde (1766-1828); 178

SUBJECT INDEX

Absolute temperature; 151
Absorption of radiation; 230, 231, 252; - spectra; 179
Acceleration; 17, 20, 114, 198; - of free fall; 126
Accompanying wave;121-123,164-165,198,213-184,186,189,197-200
Add a cycle game; 75, 77, 128
Add a dimension, add a particle games; 5, 6, 78, 93
Addition of vectors (forces, velocities, accelerations); 151
Addition of velocities treating the Doppler Effect; 225-228
Aggregation states; 2, 40-42; --of nuclear matter; 2, 44-45
Alkali halide crystals; 103-104, 170
Alpha particles; 1, 2, 6, 33, 84,
Amplitude of vibration(s), wave(s); 41, 151-154, 185
AMU, or amu, atomic mass unit; 32
Anderson experiment; 86-89, 98, 174
Angstrom, symbol Å, atomic length unit; 33.
Anti-gravity (fifth force); anti-matter; 58
Applicability limits, of approximation(s), law(s), models; 12, 70
Archimedes' law; 14, 18, 28

Band spectrum of light; 179
Bending of light; 101, 234-236
Bicycle stability syndrome; 58-59, 62, 248, 256
Big Bang; 47, 58, 64, 240, 245, 247, 253, 254
Binding energy; in atom, molecule; 38, 189
  -- in lattice; 92, 151, 155, 170
Blackbody radiation; 184; - "background" (microwave) radiation; 240
Black holes; 2, 243, 244, 257, 258
Blueshift of radiation; 223, 234
Bohr's postulates for stability of atomic orbits; 194, 196
Bohr-de Broglie condition; 196-198
Bulk deformation waves; 153-157
Buoyancy, buoyant force; 13-14, 18-19

Calory, diet(icians), unit of energy; 34
Canal rays; 1, 224; Cathode rays; 1
Centripetal acceleration; 17, 20, 199
Channel of accompanying (space-) wave; 122-123
Coherent waves; 180, 202-203
Collapse: gravitational; 255-256; of epola; 256-258
Conditions (of gas): critical; 40; normal; 40
Continuous: matter, media; 2, 46; - radiation spectra; 178
Coordinate(s), - axes; 95, 96
Copernican model of planetary motions; 69-70
Corpuscles, corpuscular theory of light; 3, 66, 162, 181, 182
Cosmic matters; dark, gray; 2, 45; - dust; 252; - rays; 1
Coulomb's law; 104, 134
Coulomb, symbol C, SI unit of electric charge; 104
Covalent binding; 103
Creation and annihilation of matter (alleged); 5, 8, 87-89, 174-176
Cross section(s); 9, 29
Crystals, crystalline phase (or state); 40

Dark matter (cosmic); 2, 45
de Broglie waves; 6,7,44,46,101,146,148
Density of matter; 25-27, 69; Average density; 29;
    In planets, Solar System, 31, 34; In atoms, 33, 34; In nuclei, 32
Diffraction, of light; 179-180; -of particle beams; 101, 140;
    -of waves; 179-180; Diffraction Grating, 180
Dilation, of length, of time (alleged); 5, 95-96
Dimensions; 5, 95-96
Discreteness of atomic matter; 3, 85-86
Dispersion of waves; 177, 233; Dispersion Spectra; 177
Doppler effect; 223-225

Earnshaw's theorem; 105, 143
Earth's mass; 31, 127;-radii, 12, 31; -velocity, 3, 31, 81
Effective mass; 7, 211, 213, 215, 217; - weight; 18
Electromagnetism; 4, 66, 139
Electromagnetic interaction; 113, 131-133
Electromagnetic theory of light; 66,183-184; - radiation 91,184,203
Electro-magneto-gravitational (EMG) field; 7,66,148-150
Electron(s); 1-3, 6-7, 89, 109
Electron-positron lattice (epola); 7, 44, 68, 107-111

262

Electron-positron pair(s); 44, 78, 86, 88-89, 92
Electron-volt (symbol eV), unit of energy; 10, 34
Electrostatic (coulombic) interaction, force(s);113,134-138,190
Electroweak interaction; 114
Emission Spectra; 179
Energy Conservation Law; 97; Energy units; 34
Epicycle(s); hypocycles; 74
Epola (electron positron lattice); collapse,256,257; temperature,241
Ether (imaginary); 4,7,66; ether currents, winds; 6,66,182,183,218
Excess particles in Half-Wave Clusters; 156
Expanding, and exploding universe (allegedly); 227, 240-241, 253-255

Face-centered cubic (symbol fcc) lattice; 101-104, 170
False vacuum bubbles (imaginary); 58, 94
fcc, see face-centered cubic
Fermi (femtometer, symbol fm) nucleic length unit; 32
Fields of forces; 4, 7, 66, 91, 105, 148-150
Field-Temperature Redshift; 241-243, 252
Fifth dimension (Kaluza's, imaginary); 5
Fifth force (anti-gravity, imaginary); 5, 58
"Fine structure constant" (just the 137 ratio of velocities); 191
Fluids (liquids *and* gases); 42
Fraunhofer lines in spectra; 178-179
Free electron, positron (free to move in space); 86-88, 211
'free' electron (free only within a solid or liquid); 36,42,162,186
Free Path of particle, molecule, atom; 167
Frequency of radiation, vibrations, waves; 121, 124, 181, 184
Frequency invariance; 221-222
Four horse rule (Aristotle's); 71, 82
Fundamental interaction(s); 113

Gamma rays; 86, 87, 119, 172, 187, 233
Gases; 40-42; 151, 168
Gate to Unit Cube of fcc lattice; 110, 216
Generalization on everything,everywhere,forever;12,45,83,97,89,209
Geocentric model of the world; 44, 72
Graininess and discreteness of atomic matter; 3, 85-86
Gravitation, gravity; 3,17,19,125
Gravitational Constant;114; - forces, interaction; 7,17-18,113,
    125-129,135-136; - mass; 19

Gravitational Collapse, 255,256; - Redshift, 235-238,248,252
Gray Matter (cosmic); 2, 45
Ground State, of atoms, - - energy; 35, 38; - - orbit; 191

Half-Wave Clusters (of Bulk Deformation Waves); 154
Halley Comet's, largest velocity in Solar System; 191
Heliocentric model of the world, 73, 75
Helium, atoms; 1,2; - nuclei, see alpha particles; spectra, 178,181
Hertz, frequency unit, symbol Hz; 121
Horoscopes (providing funds for science); 74, 77
Hubble-Humason redshifts (physically proportional to distance); 245
    their physical explanation; 251-253
    their twisted interpretation by runaway of the emitters; 246,249
Hubble "constant" (five-fold variable) 249, 251
Huygens' (wave) theory of light; 4, 182
Hydrogen atom; 1,30,33,35; -- spectrum; 181-196
Hypocycles, hypo-epicycles; 74
Hypotheses(meant EXPLANATIONS) *non fingo*; 66

Ideal gas; 40, 167; Ideal geometric shapes; 9, 11, 12
Ignoramus vs Ignorabimus; 61, 63-64, 150
Imaginary particles; 5, 47, 93
Imperfections, Impurities; in crystals, 153; in epola, 243-244, 252
Inertia; 3, 7, 19, 56, 61, 92
Inertial forces (real); 18; - interaction; 118-120, 133, 135
Inertial interaction channel (also spacewave channel); 123, 124
Inertial mass, 20; - resistance to acceleration; 212
Infrared radiation; 178
Interaction-carrying space (epola); 116
Interference of waves; 154, 163; of light; 179, 180
Interference of particles, due to the interference of their
    accompanying waves; 163
Interference spectra 179, 180
Intrinsic Magnetic Moments; 141, 142
Ionic binding; 103; ions; 101, 102, 109-112; ionization; 189

Joule, or watt·second, energy unit, symbol J; 34

Kelvin, unit of temperature in absolute scale, symbol K; 151
Kepler's model of planetary motions; 76-77
Kilo, metric prefix for thousand-fold multiplication of units; 11
Kilogram (kg), unfit name for SI unit of mass (see lav); 17, 22-23;
Kilogram-force, kgf, popular (non-SI) unit of force; 17,18,23
Kilowatt·hour, kWh, practical (non-SI) energy unit; 34
Kinetic energy; 21, 167, 190

Lattice; 6,7; - constant, - points, - sites, - units;101-104,170
Lav, proposed name for the SI unit of mass; 23, 126
Lavoisier law of mass conservation; 20-21, 176, 210
Law of gravitation (Newton's); 125, 127,128
Laws of motion (Newton's); first; 3,7; second; 20; third; 115
Lifetime; of particles, 44; of waves, 153
Light-(or radiation-) carrying (*luminiferous*) space; 66, 82, 183
Liquid droplet model of atom; 46, 84, 97
Line spectra, of absorption, emission; 178, 179
Lithium; 27, 43
Longitudinal waves; 154, 171

Madelung constant; 105, 109
Magnetic moments, interactions; of electric currents, 132,138-139;
    -- , - , of nuclear particles; 140, 141-142
Malta yok, "there is no Malta" symptom; 62, 78
Mass Conservation and indestructibility, see Lavoisier law
Mass of Earth; 31; ---calculation, 127
Mass-energy relation (alleged equivalence); 21,46,87-89,97,173-176
Mass-velocity dependence; 45, 208
Material point; 53, 161
Maxwell's electromagnetic theory of radiation;4,66,183-184,192,219
Mean free path of atom, molecule, particle; 40
Mean thermal energy of atom, molecule, particle; 35
Meson, family of unstable nuclear particles; 1
Metallic binding; 103
Metric system; 10; --prefixes; 10-11; --units; 10, 14-15
Michelson-Morley experiments; 4-6, 66, 81, 82, 183, 248
Micron, popular name for the micrometer, μm, unit of length; 9, 10

Momentum; of accompanying wave, 215-216; of molecules, 167;
- of photons, phonons, 162, 163
Monolayer; of ions, 156; of molecules, particles, 156
Muon(s), family of unstable particles; 257

Natural perceptions, concepts; 2, 46, 78
Natural philosophy; 18, 56; Natural physics; 7-8, 55-56
Natural system of orientation; 95-96
Neptune, planet; data: 31; discovery: 127
Net; 6, 7, 43
Neutrino; 44, 86; Neutron(s); 1, 30, 37, 44, 78
New physics; 5, 8
Newton laws, see law of gravitation, laws of motion
newton, SI unit of force, symbol N; 17, 126
Newton's corpuscular theory of light; 4, 182
Normal conditions of a gas; 40, 167
Nuclear forces, interactions; 114; range, 115
Nuclear matter; 1-2, 44; -particles; 1, 31; -stars; 1, 45; dust; 1, 252
Nuclei of atoms; 1-3, 6, 84

Orbit Adjustment Redshift; 239, 252
Orbital (orbiting) electron(s) in atom(s); 2, 35-37, 84
Orbital energy balance; 189-191; - velocity, 191
Orbital de Broglie wave, spacewave; 146, 197-199
Orbital magnetic moment of atomic electrons; 139
Orbital radii; 189; orbit stability; 198-201
Outer orbital (valence) electrons; 35-37, 42

Particle-wave duality (alleged); 163-165
Parsec, astronomical unit of distance,=3.3 "light year" distances; 246
Pauli force; 146-147; - (exclusion) principle; 204-205
Penetrability and Non-penetrability of atoms; 36, 37
Period of vibrations, of wave motion; 121
Perpetuum mobile; 143
Phase of substance (crystalline, superconductive, etc); 40
Phase, of vibration, wave; 154
Phase transition(s) in substances; 41
Phonon(s); 160-163, 165;
Photoelectric effect; 184-186; Photoionization, 162, 187
Photon(s); 162, 186-187; of visible light, 185

Photon of accompanying wave, its energy, mass, momentum; 213-215
Planck's constant, law; 158-159,162,185; postulate; 184-185
Planets; 3, 29-31, 73, 127; Planetary structure of atoms; 85
Pluto, planet; data: 31; discovery of: 127
Point charge (electric); 134, 161
Polarization of light, radiation; 183, 203
Polycrystalline phase; 170, 171
Positron; 1, 6, 7, 92
Postulates, postulatory or axiomatic sciences; 5, 7, 61, 93
Powers of ten notation of numbers; 10-11
Principles of Natural Physics; 55-56
Proton(s); 1, 30, 34
Ptolemaic model of the world; 47, 74-77, 78, 128
Pythagoras theorem; 216

Quanta of wave energy, sound, light, other radiation; 87, 157-159, 184; see also phonons, photons;
Quantity of matter, and volume; 12-13; and weight; 17-18; and mass; 19-22
Quantum or wave mechanics; 67-68, 200
Quantum numbers for orbital electrons in atoms; 200, 201
Quantum theory; 6-7, 63, 67, 205

Radar, 225; Radio waves; 172, 183 (Hertzian), 225
Radii of atoms; 33, 192; of nuclear particles, nuclei, 32;
 - of Earth; 12; of planets; 31
Range(s) of interaction(s); 114-115
Redshifts in radiation spectra; 223, 229, 235, 239, 241, 246, 252
Reflection of waves; 222-228
Refraction, refractive index; 177, 182, 221, 228
Relativity theory; 7, 21, 62, 67-68, 82-83, 87-88, 94
Reversed Rutherford experiment (proposed); 43
Runaway galaxies (alleged); 246, 247, 253
Rutherford experiments; 2, 6, 33, 84-86
Rutherford's planetary model of the atom; 192-193

Scales, pan and spring; 22, 91
Shapes of bodies; 9-11
Shell model of electrostatic interaction: in ionic crystals; 105
　---- in the epola space; 137-138
Short-range repulsion: in ionic crystals; 106-107, 144-145;
　--- in atomic matter, 155; in the epola; 108, 146
SI- *Systéme International* of units; 17, 23, 26, 28
Single Crystals; 170, 171
Sizes of bodies; 9-10, 49; of atoms; 33, 192; of planets; 31
Sodium chloride (rocksalt, table salt); 101-107,109,111,169-172,174
Solar System; 2, 29, 30, 73; data: 31
Solids, solid state; 40-41
Spacewave(s), see accompanying or de Broglie waves of matter
Specific Heat; 151
Spectrum of radiation; 177; Spectroscope, spectroscopy, 178
Spherical shape approximation; 11-12, 29, 109
Spin; 140; spin magnetic moments; 140, 141, 201
Standing wave(s); 198-199, 211-212
STM - Scanning tunneling microscopy; 33
Superlumic motion of nuclear particles; 218
Superposition of waves, superposition principle; 201, 202
System of orientation; 95-96; -of units, see SI

Takeover of physics; 4-5, 7-8, 51-52, 54-55, 56
Thermal energy; 35, 151; - velocity; 41; motion, vibrations; 151, 153
Time dilation, reversal; 5, 94
Tools of physics; 6, 8, 53, 55
Transverse waves; 154, 171, 172
Twin effect (alleged); 94

Ultraviolet radiation; 88, 178
Unit cube of fcc lattice; 102-103, 110
Unstable nuclei,- nuclear particles; 1, 44, 256-257
Uranium 238 atom; 30, 33, 35; -- nucleus; 30, 43, 44

Valence electrons of atoms (outermost, easiest to free or capture); 42
Vectors, vector addition; 151; Velocity addition; 227
Velocity of light; 7, 83, 172, 183, 225, 230-232
Velocity limits of atomic bodies;207-208-194; of nuclear particles;218
Velocity of sound; 169-171, 225
Vibration(s); Vibrational energy and amplitude; 41, 121, 151

Wave(s); 4, 140; bulk deformation waves; 140; velocity; 141
Wave energy (or intensity), *vs.* quantum energy; 143, 145, 203
Wave guide; 122; Wavelength; 121-123
Wavelengths and frequencies of visible light; 181
Wave-packet or wave train; 122
Wave theory of light; 4, 66, 182
"Waves of matter", 6; see Accompanying or de Broglie or space waves
Weight; 17, 18; Weightlessness; 18
Winds in ether (alleged); 4, 82
Work function ("exit toll" energy of 'free' electron); 186

X-rays, propagation velocity; 101, 162, 184, 187, 233

BIBLIOGRAPHY

1. S.Borowitz and A.Beiser, *Essentials of Physics*, Addison Wesley, Reading, Massachusets, 1967
2. C.E.Swartz, *Phenomenal Physics*, J.Wiley, New York 1981
3. C.Kittel, *Introduction to Solid State Physics*, J.Wiley, New York 1976
4. N.F.Mott and R.W.Guerney, *Electronic Processes in Ionic Crystals*, Oxford University Press, 1946
5. Max von Laue, *Geschichte der Physik*, Athenaeum, Bonn 1950
6. F.Hoyle and J.Narlikar, *The Physics-Astronomy Frontier*, W.H.Freeman, San Francisco 1980
7. D.E.Gray (Editor), *American Institute of Physics Handbook*, McGraw Hill, New York 1972
8. Asimov's *Biographical Encyclopedia of Science and Technology*, Doubleday, New York 1982
9. M.Simhony, *The Electron Positron Lattice Space*, Physics Section 5, The Hebrew University, Jerusalem 1990
10. W.Perrett and G.B.Geffery, *The Principle of Relativity*, Dover Publications, 1923

Albert Einstein's statements were extracted from his articles, (published 1905-1917, in German), and are quoted from their English translation in the book by Perrett and Geffery (Reference 10).